Problems and Solutions in Medical Physics

Series in Medical Physics and Biomedical Engineering

Series Editors: John G. Webster, E. Russell Ritenour, Slavik Tabakov,
and Kwan Hoong Ng

Recent books in the series:

Clinical Radiotherapy Physics with MATLAB: A Problem-Solving Approach
Pavel Dvorak

Advances in Particle Therapy: A Multidisciplinary Approach
Manjit Dosanjh and Jacques Bernier (Eds)

Radiotherapy and Clinical Radiobiology of Head and Neck Cancer
Loredana G. Marcu, Iuliana Toma-Dasu, Alexandru Dasu, and Claes Mercke

Problems and Solutions in Medical Physics: Diagnostic Imaging Physics
Kwan Hoong Ng, Jeannie Hsiu Ding Wong, and Geoffrey D. Clarke

Advanced and Emerging Technologies in Radiation Oncology Physics
Siyong Kim and John W. Wong (Eds)

A Guide to Outcome Modeling In Radiotherapy and Oncology: Listening to the Data
Issam El Naqa (Ed)

Advanced MR Neuroimaging: From Theory to Clinical Practice
Ioannis Tsougos

Quantitative MRI of the Brain: Principles of Physical Measurement, Second edition
Mara Cercignani, Nicholas G. Dowell, and Paul S. Tofts (Eds)

A Brief Survey of Quantitative EEG
Kaushik Majumdar

Handbook of X-ray Imaging: Physics and Technology
Paolo Russo (Ed)

Graphics Processing Unit-Based High Performance Computing in Radiation Therapy
Xun Jia and Steve B. Jiang (Eds)

Targeted Muscle Reinnervation: A Neural Interface for Artificial Limbs
Todd A. Kuiken, Aimee E. Schultz Feuser, and Ann K. Barlow (Eds)

Emerging Technologies in Brachytherapy
William Y. Song, Kari Tanderup, and Bradley Pieters (Eds)

Environmental Radioactivity and Emergency Preparedness
Mats Isaksson and Christopher L. Rääf

The Practice of Internal Dosimetry in Nuclear Medicine
Michael G. Stabin

Problems and Solutions
in Medical Physics
Diagnostic Imaging Physics

Kwan Hoong Ng
Jeannie Hsiu Ding Wong
Geoffrey D. Clarke

CRC Press
Taylor & Francis Group
Boca Raton London New York

CRC Press is an imprint of the
Taylor & Francis Group, an **informa** business

CRC Press
Taylor & Francis Group
6000 Broken Sound Parkway NW, Suite 300
Boca Raton, FL 33487-2742

International Standard Book Number-13: 978-1-4822-3995-9 (Paperback)

International Standard Book Number-13: 978-1-138-54258-7 (Hardback)

Library of Congress Cataloging-in-Publication Data

Names: Ng, Kwan Hoong. | Wong, Jeannie Hsiu Ding. | Clarke,
Geoffrey David, 1956–.
Title: Problems and solutions in medical physics.
Description: Boca Raton, FL : CRC Press,Taylor & Francis Group, [2018]- |
Series: Series in medical physics and biomedical engineering | Includes
bibliographical references.
Identifiers: LCCN 2018003314 (print) | LCCN 2018017752 (ebook) |
ISBN 9781351006781 (eBook) | ISBN 9781351006774 (eBook Adobe Reader) |
ISBN 9781351006767 (eBook ePub) | ISBN 9781351006750 (eBook Mobipocket) |
ISBN 9781482239959 | ISBN 9781482239959 (pbk; alk. paper) |
ISBN 9781138542587 (hardback; alk. Paper)
Subjects: LCSH: Medical physics—Problems, exercises, etc. | Diagnostic
imaging—Problems, exercises, etc.
Classification: LCC R896 (ebook) | LCC R896 .P76 2018 (print) | DDC 610.1/53076—dc23
LC record available at https://lccn.loc.gov/2018003314

Visit the Taylor & Francis Web site at
http://www.taylorandfrancis.com

and the CRC Press Web site at
http://www.crcpress.com

Visit the eResources:
https://crcpress.com/9781482239959

Contents

About the Series

The *Series in Medical Physics and Biomedical Engineering* describes the applications of physical sciences, engineering, and mathematics in medicine and clinical research.

The series seeks (but is not restricted to) publications in the following topics:

- Artificial organs
- Assistive technology
- Bioinformatics
- Bioinstrumentation
- Biomaterials
- Biomechanics
- Biomedical engineering
- Clinical engineering
- Imaging
- Implants
- Medical computing and mathematics
- Medical/surgical devices

- Patient monitoring
- Physiological measurement
- Prosthetics
- Radiation protection, health physics, and dosimetry
- Regulatory issues
- Rehabilitation engineering
- Sports medicine
- Systems physiology
- Telemedicine
- Tissue engineering
- Treatment

The *Series in Medical Physics and Biomedical Engineering* is an international series that meets the need for up-to-date texts in this rapidly developing field. Books in the series range in level from introductory graduate textbooks and practical handbooks to more advanced expositions of current research.

The *Series in Medical Physics and Biomedical Engineering* is the official book series of the International Organization for Medical Physics.

THE INTERNATIONAL ORGANIZATION FOR MEDICAL PHYSICS

The International Organization for Medical Physics (IOMP) represents over 18,000 medical physicists worldwide and has a membership of 80 national and 6 regional organizations, together with a number of corporate members. Individual medical physicists of all national member organisations are also automatically members.

The mission of IOMP is to advance medical physics practice worldwide by disseminating scientific and technical information, fostering the educational and professional development of medical physics and promoting the highest quality medical physics services for patients.

A World Congress on Medical Physics and Biomedical Engineering is held every three years in cooperation with International Federation for Medical and Biological Engineering (IFMBE) and International Union for Physics and Engineering Sciences in Medicine (IUPESM). A regionally based international conference, the International Congress of Medical Physics (ICMP) is held between world congresses. IOMP also sponsors international conferences, workshops and courses.

The IOMP has several programmes to assist medical physicists in developing countries. The joint IOMP Library Programme supports 75 active libraries in 43 developing countries, and the Used Equipment Programme coordinates equipment donations. The Travel Assistance Programme provides a limited number of grants to enable physicists to attend the world congresses.

IOMP co-sponsors the *Journal of Applied Clinical Medical Physics*. The IOMP publishes, twice a year, an electronic bulletin, *Medical Physics World*. IOMP also publishes e-Zine, an electronic news letter about six times a year. IOMP has an agreement with Taylor & Francis for the publication of the *Medical Physics and Biomedical Engineering* series of textbooks. IOMP members receive a discount.

IOMP collaborates with international organizations, such as the World Health Organisations (WHO), the International Atomic Energy Agency (IAEA) and other international professional bodies such as the International Radiation Protection Association (IRPA) and the International Commission on Radiological Protection (ICRP), to promote the development of medical physics and the safe use of radiation and medical devices.

Guidance on education, training and professional development of medical physicists is issued by IOMP, which is collaborating with other professional organizations in development of a professional certification system for medical physicists that can be implemented on a global basis.

The IOMP website (www.iomp.org) contains information on all the activities of the IOMP, policy statements 1 and 2 and the 'IOMP: Review and Way Forward' which outlines all the activities of IOMP and plans for the future.

Preface

In view of the increasing number and popularity of master's and higher level training programmes in medical physics worldwide, there is an increasing need for students to develop problem-solving skills in order to grasp the complex concepts which are part of ongoing clinical and scientific practice. The purpose of this book is, therefore, to provide students with the opportunity to learn and develop these skills.

This book serves as a study guide and revision tool for postgraduate students sitting examinations in radiological physics or diagnostic imaging physics. The detailed problems and solutions included in the book cover a wide spectrum of topics, following the typical syllabi used by universities on these courses worldwide.

The problems serve to illustrate and augment the underlying theory, and provide a reinforcement of basic principles to enhance learning and information retention. No book can claim to cover all topics exhaustively, but additional problems and solutions will be made available periodically on the publisher's website, https://crcpress.com/9781482239959.

One hundred and thirty-three solved problems are provided in eleven chapters on basic physics topics, including: X-ray generation, screen-film radiography, digital radiography, imaging, mammography, fluoroscopy, computed tomography, magnetic resonance imaging, ultrasound, radiobiology and radiation protection. The approach to the problems and solutions covers all six levels in the cognitive domain of Bloom's taxonomy.

This book is one of a three volume set containing medical physics problems and solutions. The other two books in the set tackle nuclear medicine physics and radiotherapy physics.

We would like to thank the staff at Taylor & Francis, especially Francesca McGowan and Rebecca Davies, for their unfailing support.

Kwan Hoong Ng, Jeannie Hsiu Ding Wong and Geoffrey D. Clarke

Authors

Kwan Hoong Ng, PhD, FinstP, DABMP, received his MSc (Medical Physics) from the University of Aberdeen and PhD (Medical Physics) from the University of Malaya, Malaysia. He is certified by the American Board of Medical Physicists. Professor Ng was honoured as one of the top 50 medical physicists in the world by the International Organization of Medical Physics (IOMP) in 2013. He also received the International Day of Medical Physics Award in 2016. He has authored/co-authored over 230 papers in peer-reviewed journals, 25 book chapters and co-edited 5 books. He has presented over 500 scientific papers and more than 300 invited lectures. He has also organized and directed several workshops on radiology quality assurance, digital imaging and scientific writing. He has directed research initiatives in breast imaging, intervention radiology, radiological safety and radiation dosimetry. Professor Ng serves as a consultant for the International Atomic Energy Agency (IAEA) and is a member of the International Advisory Committee of the World Health Organization, in addition to previously serving as a consulting expert for the International Commission on Non-Ionizing Radiation Protection (ICNIRP). He is the founding and emeritus president of the South East Asian Federation of Medical Physics (SEAFOMP) and is a past president of the Asia-Oceania Federation of Organizations for Medical Physics (AFOMP).

Jeannie Hsiu Ding Wong, PhD, is a senior lecturer at the Department of Biomedical Imaging, Faculty of Medicine, University of Malaya, Kuala Lumpur, Malaysia. She coordinated the Master of Medical Physics programme from 2013 to 2017. Dr. Wong received her Bachelor's degree in Biomedical Engineering from University of Malaya in the year 2003. In 2004, Dr. Wong obtained her Master of Medical Physics degree from the University of Malaya. In 2008, she furthered her studies at the University of Wollongong, Australia. She earned her PhD in 2011. Dr. Wong's research interests focus on radiation physics and radiation dosimetry. She has published more than 32 peer-reviewed articles, a book chapter, 7 conference proceedings and more than 40 scientific papers in both local and international conferences to date.

Geoffrey D. Clarke, PhD, FACR, FAAPM, is a professor of Radiology at the University of Texas Health Science Center at San Antonio (UTSHCSA), Texas. He is also the director of the Graduate Programme in Radiological Science and the chief of the MRI Division for the Research Imaging Institute at UTSHCSA. He has served in leadership positions on boards and committees for various societies, including the AAPM, CAMPEP, ACR, ABMP and ACMP. Dr. Clarke was one of the first generation of MRI scientists to use the technology for investigating biomedical problems. Early on, he developed imaging technologies and spectroscopic methods to study head trauma and coronary artery disease. He has received research grants from the National Institutes of Health and American Heart Association. His current research includes evaluating skeletal muscle metabolism in diabetes and measuring impaired cardiac function due to perinatal stresses using magnetic resonance. As leader of the Imaging Core Research Laboratory at the Southwest National Primate Center, Dr. Clarke also provides technical support to colleagues who seek to develop practical image acquisition and analysis methods for their research programs.

Acknowledgements

We acknowledge the contribution from the following people:

Azlan Che Ahmad
Department of Biomedical Imaging
Faculty of Medicine
University of Malaya
Kuala Lumpur, Malaysia

Ying Ying Cheah
Alypz Sdn. Bhd.
Selangor, Malaysia

Muhammad Shahrun Nizam Daman Huri
Department of Biomedical Imaging
Faculty of Medicine
University of Malaya
Kuala Lumpur, Malaysia

Yen Ling Leong
Catholic High School
Ministry of Education
Petaling Jaya, Malaysia

Khadijah Ramli
Department of Biomedical Imaging
Faculty of Medicine
University of Malaya
Kuala Lumpur, Malaysia

Li Kuo Tan
Department of Biomedical Imaging
Faculty of Medicine
University of Malaya
Kuala Lumpur, Malaysia

Basic Physics

1

1.1 TRANSMISSION INTENSITY

Problem:
Refer to the attenuation curves (Figure 1.1), explain the difference between 60 keV gamma ray and 60 kVp X-ray beam.

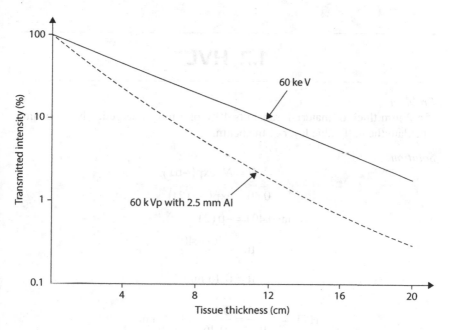

FIGURE 1.1   Attenuation in soft tissue for a 60 keV gamma ray and for a 60 kVp X-ray beam.

Solution:

The attenuation is the reduction in the number of photons in the beam and for photons of uniform energy (i.e. monoenergetic such as 60 keV gamma ray) it follows the simple exponential law.

$$N = N_o \exp(-\mu x)$$

Where,

N = number of photons transmitted

N_o = initial number of photons

μ = linear attenuation coefficient (cm^{-1})

x = thickness of material (cm)

For polyenergetic radiation, such as the 60 kVp X-ray beam, the effect of beam hardening where the lower energy photons are preferentially removed from the beam effectively reduces the slope of the curve (log intensity vs. thickness) and a pure linear relationship is no longer valid.

1.2 HVL

Problem:

If a 2 mm thick of material transmits 40% of a monoenergetic photon beam, calculate the half value layer of the beam.

Solution:

$$N = N_o \exp(-\mu x)$$
$$0.40 = \exp(-\mu(2))$$
$$\ln(0.40) = -\mu(2)$$
$$\mu = -\frac{\ln(0.40)}{2}$$
$$\mu = 0.46 \text{ mm}^{-1}$$

$$\text{HVL} = \frac{0.693}{\mu} = \frac{0.693}{0.46} = 1.51 \text{ mm}$$

1.3 K-EDGE

Problem:

    a. Figure 1.2 shows a mass attenuation coefficient curve. Name the peak *A* and suggest an element that fits this curve.

    b. Explain the physics phenomenon exhibited by *A*.

    c. Give an example of a clinical application of this phenomenon.

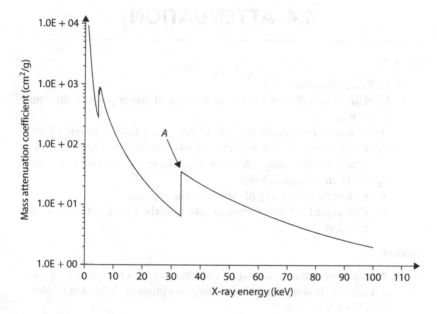

FIGURE 1.2 Mass attenuation coefficient curve.

Solution:

    a. Peak *A* is the K-edge (K-absorption edge) for iodine. Its energy is 33.2 keV.

    b. K-edge is the binding energy of the K-shell electron of an atom. There is a sudden increase in the attenuation coefficient of photons just above the binding energy of the K-shell electron (binding energy of K-shell for iodine = 33.2 keV).

      This sharp increase in attenuation is due to photoelectric absorption effect. For this interaction to occur, the photons must have slightly more energy than the K-edge. Therefore, a photon with energy just above the binding energy of the electron is more likely to be absorbed than a photon with energy below this binding energy.

c. Iodine is used as a contrast medium (or agent) in enhancing the contrast of blood vessels during X-ray imaging since the K-edge of iodine is 33.2 keV, which is close to the mean energy of common diagnostic X-ray beam (e.g. the mean energy of a 100 kVp X-ray beam is between 33 and 50 keV).

1.4 ATTENUATION

Problem:

a. Define attenuation.
b. Explain how radiation energy is attenuated while passing through a medium.
c. A beam of X-rays, consisting of 10^9 photons, passes through a 5 cm thick slab (medium) where there are two attenuation processes taking place. Using linear attenuation coefficient $\mu_1 = 0.02$ cm^{-1} and $\mu_2 = 0.04$ cm^{-1}, respectively:
   i. Calculate how many photons are transmitted.
   ii. Calculate how many photons are absorbed by each process in the slab.

Solution:

a. Attenuation is the reduction of the intensity of a radiation beam as it traverses through matter by either absorption or deflection of photons from the beam.
b. When a photon beam passes through matter, some of its energy is absorbed (energy is transferred to matter). Some of the photons are scattered, i.e. they collide with outer shell electrons in atoms and lose some energy and are forced to change direction (Figure 1.3).
c. Calculation
   i. Number of particles transmitted,

   $$N_T = N_o e^{-(\mu_1 + \mu_2)L}$$
   $$= 10^9 e^{-(0.02+0.04)5}$$
   $$= 7.41 \times 10^8$$

   ii. Total number of particles absorbed,
   $$N_o - N_T = \left(10^9 - 7.41 \times 10^8\right) = 2.59 \times 10^8$$

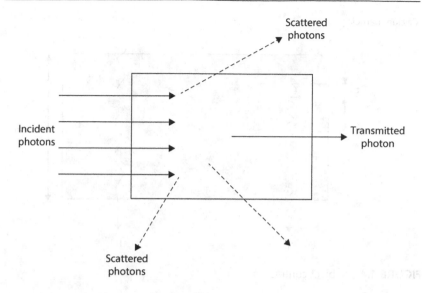

FIGURE 1.3 Attenuation of radiation.

Number of particles absorbed by process 1,

$$N_1 = (N_o - N_T)\frac{\mu_1}{\mu_1 + \mu_2} = 2.59 \times 10^8 \times \frac{0.02}{0.06} = 8.63 \times 10^7$$

Number of particles absorbed by process 2,

$$N_2 = (N_o - N_T)\frac{\mu_2}{\mu_1 + \mu_2} = 2.59 \times 10^8 \times \frac{0.04}{0.06} = 1.73 \times 10^8$$

1.5 CONTRAST

Problem:

a. Refer to the Figure 1.4. A beam of X-ray photons incident upon a slab of tissue with a thickness of 1 mm embedded with a calcium particle with a thickness, $t = 100$ μm. Calculate the percentage subject contrast between the calcium particle and tissue for 20, 50 and 100 keV, respectively. Assume there is no scatter. The linear attenuation coefficient values are given in Table 1.1.

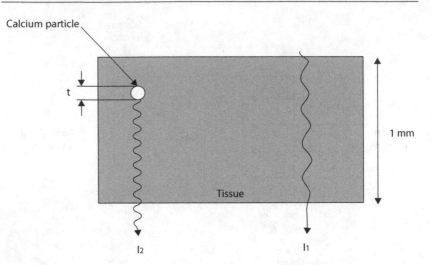

FIGURE 1.4   Subject contrast.

TABLE 1.1   Linear attenuation coefficients

keV	TISSUE, μ (cm^{-1})	CALCIUM μ (cm^{-1})
20	0.793	20.15
50	0.227	1.547
100	0.170	0.397

b. Based on the results from a., explain the optimal choice of photon energy for visualisation of small calcium particles in tissue.

Solution:

a. Subject contrast,

$$C_{\text{subject}} = \log_{10} I_1 - \log_{10} I_2 = \log_{10} \frac{I_1}{I_2}$$

$$\%C_{\text{subject}} = \log_{10} \frac{I_1}{I_2} \times 100$$

At 20 keV, transmitted particles for tissue only region,

$$I_1 = I_o \exp{-\left(\mu_{\text{tissue}} \times t_{\text{tissue}}\right)}$$
$$= I_o \exp{-\left(0.793 \times 0.1\right)}$$
$$= 0.924 \, I_o$$

Transmitted particles for tissue and calcium region,

$$I_2 = I_o \exp-\left[\left(\mu_{\text{tissue}} \times t_{\text{tissue}}\right) + \left(\mu_{\text{calcium}} \times t_{\text{calcium}}\right)\right]$$
$$= I_o \exp-\left[\left(0.793 \times 0.09\right) + \left(20.15 \times 0.01\right)\right]$$
$$= 0.761 I_o$$

Hence, subject contrast at 20 keV is

$$\%C_{\text{subject}} = \log_{10} \frac{I_1}{I_2} \times 100$$
$$= \log_{10} \frac{0.924\ I_o}{0.761 I_o} \times 100$$
$$= 8.4\%$$

Results:

keV	TISSUE, μ (cm^{-1})	CALCIUM μ (cm^{-1})	$\Delta\mu$	$C_{subject}$ (%)
20	0.793	20.15	19.36	8.4
50	0.227	1.547	1.32	0.57
100	0.170	0.397	0.227	0.1

From the results, it can be observed that the subject contrast decreases with increased photon energy; therefore, optimal visualisation of small particles occurs at lower photon energy.

X-ray Production

2

2.1 X-RAY TUBE

Problem:
Label the parts of a rotating anode X-ray tube in Figure 2.1.

Solution:
A: Anode target, B: Bearing, C: Cathode filament, D: Envelope, E: Housing, F: Target, G: Stator, H: Rotor, I: Oil, J: Window

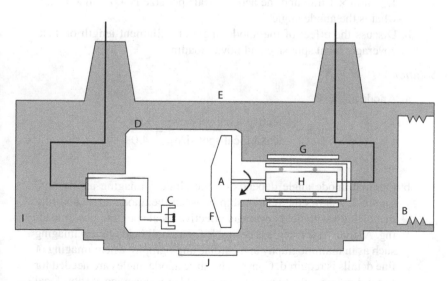

FIGURE 2.1 Rotating anode X-ray tube.

2.2 HEAT

Problem:
During the production of X-rays, not all of the energies lost by the electrons are converted to X-ray photons; some of the energies are converted to heat. What is the ratio of heat: X-ray yield for a conventional diagnostic X-ray tube operating at 120 kVp? Compare this to a therapeutic X-ray tube operating at 4 MV.

Solution:
For 120 kVp, the heat: X-ray yield is 99:1, while it is 60:40 at 4 MV.
At low tube potentials, the production of X-ray photons is very inefficient, resulting in a large proportion of the electrons' energy being converted to heat.

2.3 FOCAL SPOT SIZE

Problem:

    a. Figure 2.2 shows the actual and effective focal spot size on a standard rotating anode X-ray tube. Given the effective focal spot size is 1.04 mm × 1 mm and the actual focal spot size is 4.0 mm × 1 mm, what is the anode angle?

    b. Discuss the effect of the anode angle and filament length on field coverage, focal spot size and power loading.

Solution:

    a. Anode angle, θ

$$\sin\theta = \frac{\text{effective focal spot size}}{\text{actual focal spot size}} = \frac{1.04}{4.0}$$

$$\theta = 15.1°$$

    b. Optimal anode angle depends on the clinical imaging application. Figure 2.3 shows the relationship between the anode angle and filament length with field coverage, effective focal spot and power loading. A small anode angle is desirable for a small field of view imaging such as in mammography and cardiac angiography where imaging of fine details is required. Conversely, large anode angles are needed for general radiography to achieve large field area coverage at short focal spot-to-image distances. An X-ray tube normally has two filaments,

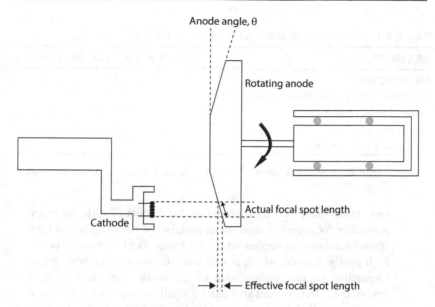

FIGURE 2.2   Focal spot size of an X-ray tube.

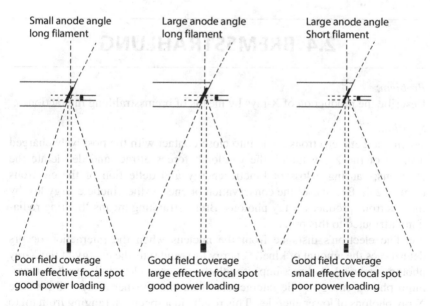

FIGURE 2.3   Effect of anode angle and filament length.

TABLE 2.1   Typical focal spot size for different modalities

MODALITY	TYPICAL FOCAL SPOT SIZE (mm)
Mammography	0.3
Mammography (with magnification)	0.1
General radiography	0.6–1.2
Fluoroscopy	0.6
Computed Tomography	1.0

Source: Allisy-Roberts, P and Williams, J. *Farr's Physics for Medical Imaging*, Saunders Ltd, Philadelphia, 2008.

one shorter than the other, providing focal spots of different sizes. A smaller focal spot is used when imaging of fine details and high spatial resolution is needed while the larger focal spot can produce high photon fluence, which is necessary to image thick body parts. Depending on the configuration of the anode angle and filament length, the effective focal spot can be small or large. A smaller focal spot size usually has a poor power loading (or higher heat ratings).

Note:
Table 2.1 shows a list of typical focal spot size for different imaging modalities.

2.4 BREMSSTRAHLUNG

Problem:
Describe the production of X-rays by means of bremsstrahlung interaction.

Solution:
When energetic electrons come into close contact with the positively charged nucleus of the X-ray target, the Coulomb forces attract and decelerate the electrons, causing a loss of kinetic energy and deflection of the electrons (Figure 2.4). Because of the conservation of energy, the kinetic energy lost by the electrons produces X-ray photons. Bremsstrahlung means 'braking radiation' attributed to this process.

The electron's distance from the nucleus when the interaction occurs determines the amount of kinetic energy lost, and thus the energy of the X-ray photon produced. A direct impact with the target nucleus will result in maximum photon energy while interactions occurring further away will produce X-ray photons of lower energies. This results in a spectrum ranging from nil to the highest energy, determined by the peak tube voltage (kVp) set.

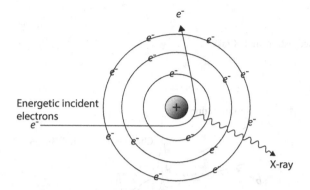

FIGURE 2.4   Bremsstrahlung interaction.

2.5 EFFECT OF kV AND mAs ON IMAGE QUALITY

Problem:
How do the tube voltage (kV) and tube current (mA) affect image quality?

Solution:
Tube voltage determines the penetrability of the beam (quality) while tube current determines the total number of photons (quantity).

The choice of kV is influenced by the organ or tissue imaged, as well as the patient's physical size. Larger patient will require higher kV compared to a smaller size patient. Imaging soft tissue, such as breast, will require low kV for higher contrast since the effective Z is low, however bone will produce good contrast at higher kV due to its higher effective Z. Increasing mA will also improve contrast but patient dose will also increase proportionally.

2.6 X-RAY SPECTRUM

Problem:
Sketch the X-ray spectrum at the following locations:

    a. After the X-ray target, before any filter
    b. After the filter, before reaching the patient

Solution:
Figure 2.5 shows the X-ray spectrum.

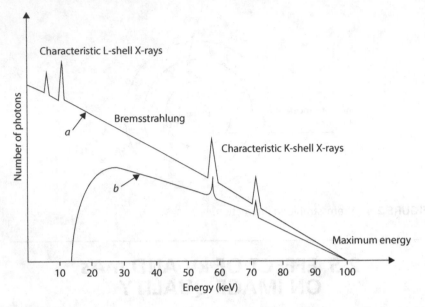

FIGURE 2.5 X-ray spectrum (a) after the X-ray target, before any filter and (b) after the filter, before reaching patient.

2.7 X-RAY OUTPUT

Problem:

How does the following X-ray tube exposure parameter affect the output of an X-ray tube?

    a. Tube voltage (kV)
    b. Tube current (mA)
    c. Time of exposure
    d. Filtration
    e. Target material

Solution:

    a. Tube voltage (kV)
       Changing tube voltage changes the total number of photons emitted (quantity) and the photon energies (quality). This is because although the number of bombarding electrons is the same, the probability of X-ray production increases with increased tube

voltage. Increasing the kV may result in the appearance of more characteristic X-ray photons and the shifting of the mean photon energy to higher energy (keV) (see Figure 2.6). The relationship

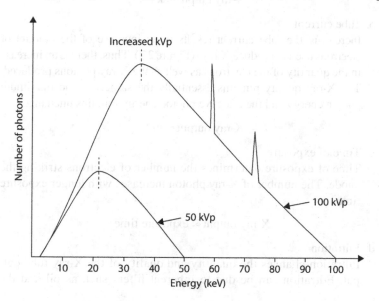

FIGURE 2.6 Effect of increasing tube voltage on X-ray output.

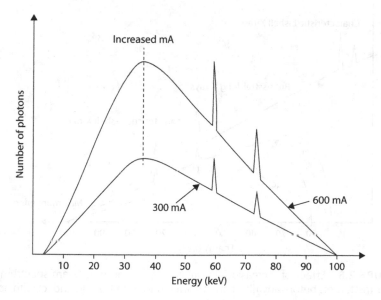

FIGURE 2.7 Effect of increasing tube current on X-ray output.

between the X-ray outputs with tube voltage is often approximated as follows:

$$\text{X-ray output} \propto kV^2$$

b. Tube current

Increasing the tube current results in an increase of the amount of electrons used to produce X-rays (Figure 2.7). Thus, there is an increase in the quantity of the electrons as well as the X-ray photons produced. The X-ray quality remains essentially the same, i.e. the maximum photon energy and the effective photon energy remains unchanged.

$$\text{X-ray output} \propto mA$$

c. Time of exposure

Time of exposure determines the number of electrons striking the anode. The number of X-ray photon increases with longer exposure time

$$\text{X-ray output} \propto \text{exposure time}$$

d. Filtration

Filtration changes the quantity and quality of the X-ray tube output. Filtration can be due to inherent filters such as oil and the

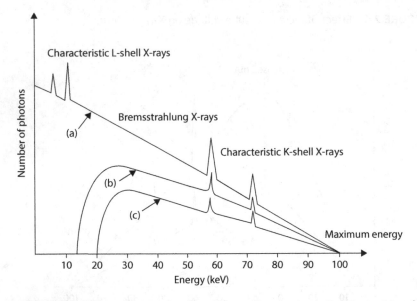

FIGURE 2.8 Effect of increasing filtration on X-ray output. Beam spectrum (a) after the target, before any filtration, (b) after inherent filtration and (c) with additional filtration.

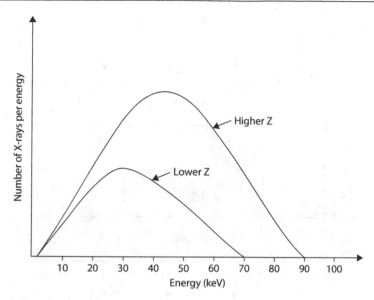

FIGURE 2.9 Effect of using different target material on X-ray output.

window of the tube housing. Filtration can also be due to additional filters placed in the beam path. With increased filtration, the number of X-ray photons is reduced and the mean energy of the X-ray energy spectrum shifts to a higher energy (Figure 2.8). This is because a proportion of the low energy photons is attenuated or filtered. K-edge filters are special filters that behave as a band-pass filter to allow the transmission of X-ray photons of a certain energy window.

e. Target material

The probability of X-ray production, or X-ray yield, increases with the use of X-ray targets made of high atomic number (Z) materials (Figure 2.9). X-ray output generally increases in the total number of photons produced. Characteristic X-ray photons with higher energies are also produced using a target material with higher Z.

2.8 OPTICAL DENSITY, CONTRAST

Problem:

Figure 2.10 shows the schematic diagram of the irradiation setup of a radiograph taken of a phantom consisting of bone, soft tissue and air gap

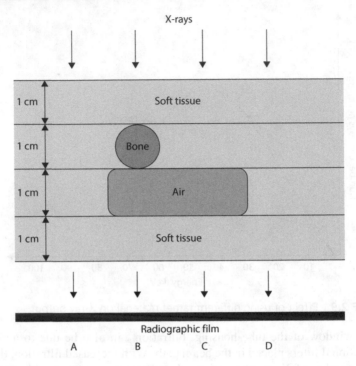

FIGURE 2.10 A phantom consisting of soft tissue, bone and air gap structures irradiated with a monoenergetic X-ray.

structures. The efficiency of the screen film system is only 30%. Assume a monoenergetic photon source and the linear attenuation coefficients of the tissues are:

bone, $\mu_{bone} = 0.5$ cm^{-1}
soft tissue, $\mu_{soft\ tissue} = 0.3$ cm^{-1}
air, $\mu_{air} = 0.0$ cm^{-1}

a. Calculate the optical density OD at positions A, B, C and D.
b. What is the contrast between position B and C? Suggest ways to improve this.

Solution:

a. At positions A and D,
Total attenuation from the X-ray photons transmitted through 4 cm of soft tissue:

$$\mu t_{A \ or \ D} = \left(\mu_{\text{soft tissue}} \times \text{thickness}_{\text{soft tissue}}\right)$$

$$= \left(0.3 \text{ cm}^{-1} \times 4 \text{ cm}\right)$$

$$= 1.2$$

Applying the Beer–Lambert formula,

$$\text{Transmitted photons}, N_t = N_o e^{-\mu t}$$

$$N_{t,A} = N_o e^{-(1.2)} = 0.301 N_o$$

Given film efficiency is 30%, hence the total number of photons absorbed by the screen film at position A or D is,

$$N_{A \ or \ D} = 0.3 \times 0.301 N_o$$

$$= 0.0903 N_o$$

Optical density is defined as

$$OD = \log_{10} \frac{N_o}{N_t}$$

Optical density at position A or D is,

$$OD_{A \ or \ D} = \log_{10} \frac{N_o}{N_{A \ or \ D}}$$

$$= \log_{10} \frac{1}{0.0903}$$

$$= 1.044$$

For positions B and C, the same formula is used, except for the total attenuation coefficients of the different tissues, which are different at positions B and C.

At position B,

$$\text{Total attenuation} = \mu t_B = \left(0.3 \text{ cm}^{-1} \times 2 \text{ cm}\right) + \left(0.5 \text{ cm}^{-1} \times 1 \text{ cm}\right)$$

$$+ \left(0.0 \text{ cm}^{-1} \times 1 \text{ cm}\right) = 1.1$$

$$N_{t,B} = N_o e^{-(1.1)} = 0.333 N_o$$

$$N_B = 0.3 \times 0.333 N_o = 0.0999 N_o$$

Optical density, $OD_B = 1.0004$

At position C,

$$\text{Total attenuation} = \mu t_B = \left(0.3 \text{ cm}^{-1} \times 3 \text{ cm}\right) + \left(0.0 \text{ cm}^{-1} \times 1 \text{ cm}\right) = 0.9$$

$$N_{t,C} = N_o e^{-(0.9)} = 0.407 N_o$$

$$N_C = 0.3 \times 0.407 N_o = 0.122 N_o$$

Optical density, $OD_C = 0.914$

b. Contrast between positions B and C

$$\text{Contrast} = \log_{10} \frac{N_{t,C}}{N_{t,B}} = \log_{10} \frac{0.407}{0.333} = 0.087$$

To improve contrast, one can decrease the tube voltage, thereby increasing the photoelectric (PE) absorption probability of the incident X-ray photons. Bony tissue, which has a higher effective atomic number, will tend to have higher PE absorption interaction and appears brighter on the film.

2.9 X-RAY SPECTRUM

Problem:
Why does X-ray beam produced from a diagnostic X-ray tube consist of photons with a broad spectrum of energies rather than a monoenergetic beam?

Solution:
X-ray photons are produced by the bremsstrahlung process whereby fast-moving electrons are slowed down by the attractive force from the nucleus. The electrons do not penetrate the nucleus but deviate from it, resulting in a loss in the kinetic energy of the electrons and the emission of electromagnetic radiation in the X-ray range of the spectrum. The amount of energy lost by such interactions is variable and hence the emitted X-ray photons have a broad spectrum of energies.

2.10 HIGH kV-SHORT EXPOSURE

Problem:
Explain why it is advantageous to increase the tube voltage when a radiograph is taken with short exposure time.

Solution:
An increase in the tube voltage results in (i) electrons that are accelerated to higher energies and production of more penetrative photons and (ii) increase in the intensity of the X-rays.

Hence, in order to produce diagnostic quality images, radiographs taken with short exposure time can increase tube voltage to increase X-ray output.

2.11 X-RAY SPECTRUM

Problem:
Plot the X-ray spectra from a tungsten X-ray tube with the peak tube potential of 120 kVp, 80 kVp, 60 kVp, and 40 kVp (Figure 2.11).
Solution:

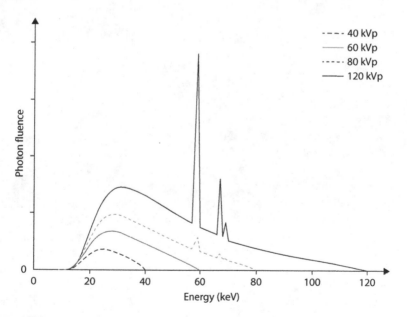

FIGURE 2.11 X-ray spectrum for a tungsten X-ray tube with peak tube potentials of 120 IVp, 80 kVp, 60 kVp and 40 kVp.

Screen Film Radiography

3

3.1 HURTER AND DRIFFIELD (H&D) CURVE

Problem:

Consider the case of an exposure taken at the 'toe', straight-line and shoulder region of the H&D curve of the film-screen system (Figure 3.1). Describe briefly what you would expect to observe on the processed film.

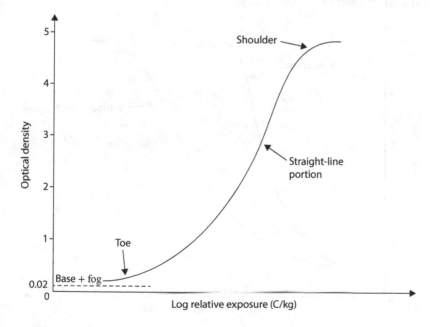

FIGURE 3.1   H&D curve.

Solution:
'Toe' region – The optical density (OD) can never be 0 even though the film was not exposed. This is 'fogging' due to the temperature, humidity and background radiation. The radiograph tends to look grainy because of high quantum noise due to the small number of photons reaching the film. Overall, the radiograph image looks bright and there is a lack of contrast throughout.

Straight line region – Image in which high contrast can be obtained.

'Shoulder' region – The region in which the image receptor detected too many photons, resulting in saturation of the film. The radiograph looks dark and there is a lack of contrast throughout.

Problem:
Draw on the same graph the H&D curves for a radiographic film used with screen, and without screen. Explain the difference between them.

Solution:
The H&D curve for a radiographic film used with screen has a higher gamma (steeper curve) and the whole curve tends to shift towards the left (Figure 3.2). This means that lower exposure is needed to produce the same film darkening and a larger contrast.

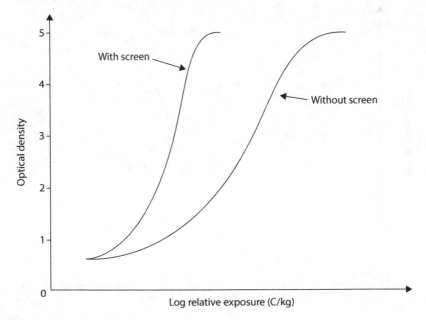

FIGURE 3.2 H&D curve for screened and unscreened film.

3.2 RADIOGRAPHIC SCREENS

Problem:
Define the intensification factor for the radiographic screen. Discuss the precautions needed when using screens.

Solution:
Intensification factor is the ratio of the exposure required for a film used without screen to film used with screen to produce the same darkening.

$$\text{Intensification factor} = \frac{\text{exposure required to produce OD 1.0 without screen}}{\text{exposure required to produce OD 1.0 with screen}}$$

Intensification factor is usually in the order of 50–100. The intensification factor is valid for one density and one kVp only.

The use of a screen results in a characteristic curve that is shifted to the left and has a steeper slope. In other words, the film has a higher gamma, enabling the use of lower exposure to produce the same darkening.
Other precautions needed when using screens include:

- Ensure that the screen film is contained in light-tight cassette to prevent fogging.
- Good compression is required to ensure a good contact between the screen and films to prevent loss of resolution.
- The front part of the cassette is made up of low-Z material (Al or plastic) so as to be radiolucent. On the other hand, the back of the cassette is made up of high-Z material to absorb X-ray photons and to prevent backscattered radiation from reaching the screen.
- Although increasing screen thickness will increase efficiency, screen thickness cannot be increased indefinitely because light photons are less energetic and tend to get absorbed by the screen material before reaching the film. Unsharpness will also occur with a thick screen because of the spreading of light photons.

3.3 FILM GAMMA AND SPEED

Problem:
Explain the meaning of the speed of a radiographic film and discuss the factors that affect it.

Solution:
Radiographic film speed is defined as

$$\text{film speed} = \frac{1}{\text{exposure in R required for an OD of 1.0 above base density}}$$

Factors that affect film speed are crystal grain size, screen thickness and energy of X-ray photons.

Films made with larger silver halide grain size require fewer X-ray photons to produce darkening of the film that is visible to the human eye. Fewer large grains need to be developed to produce the same darkening.

Screen thickness – films using thicker screens require fewer X-ray photons to produce film darkening because it is more likely that a given X-ray photon will interact with a thicker screen.

Energy of X-ray photons – essentially, the photoelectric absorption cross-section is higher for X-ray photons with lower energies. However, since there is a wide spectrum of X-ray photons produced through the bremsstrahlung process, this needs to be made more clear, especially in the context of rare-earth screens, which have K-edges especially selected for their sensitivity to X-rays produced at diagnostic kVps.

3.4 CONTRAST

Problem:
A radiograph was found to lack contrast. On the repeat radiograph, how can increasing the current (mA) produce a radiograph with higher contrast? Explain the reason for this.

Solution:
If the cause of the lack of contrast in the radiograph is due to under-exposure, increasing the current (mA) on the repeat radiograph will increase the contrast. This is because the exposure latitude of the radiograph which was initially placed at the 'toe' part of the H&D curve is now placed at the linear portion of the curve.

Problem:
Optical densities of region A and B on a radiograph are 1.0 and 1.5, respectively.

  a. Calculate the contrast between the two regions.
  b. The radiograph has a film gamma of 1.0. If an exposure of $4.0\,\mathrm{C\,kg^{-1}}$ was delivered to region A, calculate the exposure to the region B.

Solution:

a. Contrast between the two regions:

$$\text{Contrast} = \log_{10}\frac{B}{A} = \log_{10}\frac{1.5}{1.0} = 0.176$$

b.

$$\log_{10}\frac{x}{4} = 0.176$$
$$\log_{10}x - \log_{10} = 0.176$$
$$\log_{10}x = 0.176 + \log_{10}4$$
$$\log_{10}x = 0.778$$
$$x = 10^{0.778} = 6.0 \text{ Ckg}^{-1}$$

Note:

There are other definitions of contrast. For example, $\text{Contrast} = \dfrac{B-A}{A}$.

3.5 SCREEN FILM

Problem:
Good contact between intensifying screen and X-ray film is needed to produce radiographic image with high resolution. Explain why.

Solution:
When the intensifying screen and X-ray film are not in good contact, the spread of light photons leaving the screen before reaching the film will result in the loss of resolution.

3.6 GRID

Problem:
With respect to grids used in an X-ray procedure, explain the meaning of

a. Grid cutoff
b. Off-distance cutoff
c. Off-centre cutoff
d. Off-level cutoff

Solution:

    a. Grid cutoff is the loss of primary radiation caused by improper alignment of a parallel radiographic grid.

    b. Off-distance cutoff (also called axial decentring) is the use of focused grid at incorrect target to grid distance. This is relevant for focused grids.

    c. Off-centre cutoff (also called lateral decentring) occurs when X-rays parallel to the strips of a focused grid coverage at a location that is displaced laterally from the target of the X-ray tube.

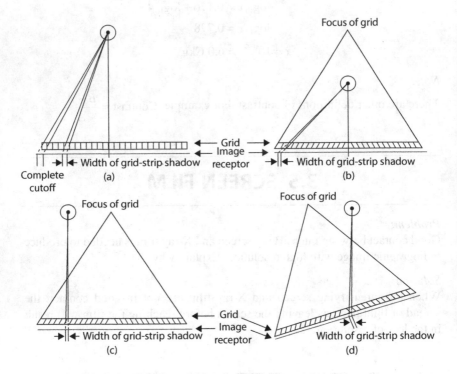

    d. Off-level cutoff results from the tilting of the focused grid, causing an overall reduction in OD across the radiograph.

3.7 GRID RATIO

Problem:

The use of grid in radiography improves image quality. Explain grid ratio and discuss the advantage and disadvantage of grid ratios of 16:1 over 12:1.

Solution:

Grid ratio is the ratio between the height of the lead strips and the distance between them. The advantage of 16:1 over 12:1 grid ratio is its more efficient removal of scattered radiations. The disadvantage 16:1 over 12:1 grid ratio is that it increases the primary radiation and therefore increased radiation exposure to patient.

3.8 RARE EARTH SCREENS

Problem:

Explain why rare-earth screens are faster than conventional calcium tungstate screens.

Solution:

Two factors that contribute to rare-earth screens being faster than conventional screens are (i) higher X-ray absorption efficiency and (ii) higher X-ray to light conversion.

Higher X-ray absorption efficiency – rare earth (40%–60%) versus $CaWO_4$ (20%–40%)

Higher X-ray to light conversion – rare earth (20%) versus $CaWO_4$ (5%)

3.9 INTENSIFYING SCREEN

Problem:

a. What is the function of intensifying screens in a cassette?
b. Name two types of rare earth phosphor used in intensifying screen. What are the advantages of rare earth phosphor over calcium tungstate?

Solution:

    a. The intensifying screen consists of phosphors that absorb X-ray photons and convert them to visible light photons that are detected and recorded by photographic film.

    b. Rare earth phosphor:

Lanthanum oxybromide (LaOBr) – blue light
Gadolinium oxysulphide (Gd$_2$O$_2$S) – green light

- Absorption edges (refer to Figure 3.3).
- Rare earth has higher conversion efficiency compared to calcium tungstate.
- Fewer X-ray photons are required to produce image, thus reducing patient dose.

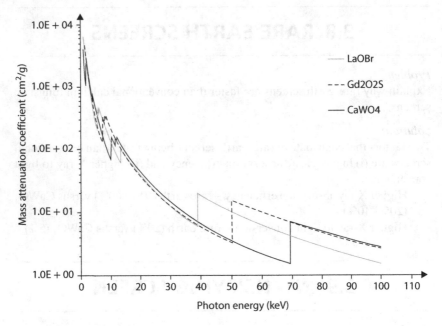

FIGURE 3.3   Absorption edge for LaOBr (39 keV), Gd$_2$O$_2$S (50.2 keV) and CaWO$_4$ (69.5 keV).

Digital Radiography

4

4.1 CHARACTERISTIC CURVES OF COMPUTED RADIOGRAPHY VS. SCREEN FILM RADIOGRAPHY

Problem:

Why are computed radiography (CR) systems considered to have better dynamic range compared to screen film (SF) systems?

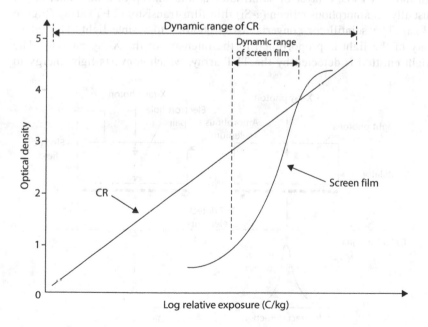

FIGURE 4.1   Characteristic curves of a CR and a SF system.

Solution:
Figure 4.1 shows the characteristic curves of a CR and a SF. An SF system has sigmoid-shaped characteristic curves. The slope of the toe and shoulder regions are flattened out (less steep) compared to the linear slope in the middle. This means that exposures made at these regions will not be able to display a change in the optical density (contrast), and hence the X-ray attenuation optimally. CR systems have a linear characteristic curve, indicating that usable contrast is achievable throughout the entire range of the exposure.

4.2 DIRECT AND INDIRECT FLAT PANEL DETECTORS

Problem:
Describe the differences between direct and indirect flat panel detectors (FPD) in terms of their detection mechanism.

Solution:
Indirect FPD
In indirect FPD, a layer of scintillator is laid on top of a flat panel array, usually an amorphous silicon (a-Si) thin-film-transistor (TFT) array (Figure 4.2a). The scintillator converts X-ray photons to visible light. The intensity of the light is proportional to the intensity of the X-ray photons. The light emitted is detected by the TFT array, which converts light energy to

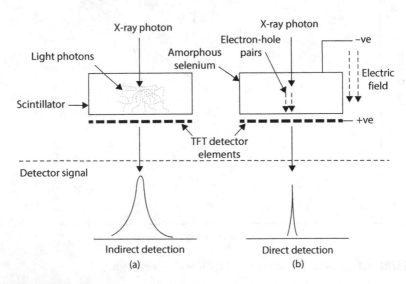

FIGURE 4.2   (a) Indirect FPD and (b) direct FPD.

electrical signal. Common scintillator materials are CsI and Gd_2O_2S. X-ray photons usually interact at the frontal surface of the scintillator material. The visible light generated has to travel a finite distance before reaching the FPD. This results in the diffusion and spreading of the light, hence causing blurring of the image. To improve on this, some scintillators are made up of columnar CsI crystals, which act as light channels directing the light signals to the FPD. This reduces spreading and thus improves spatial resolution of the system.

Direct FPD

In direct FPD, a photoconductor is used as the radiation detection material (Figure 4.2b). One of the most common photoconductors is amorphous selenium (a-Se). A-Se is layered between two electrodes and a bias voltage is applied. When X-ray photons traverse the photoconductor, they are absorbed and electron-hole pairs proportional to the X-ray intensity are generated in the solid state material. The electrons and holes are attracted to the positive and negative electrodes and converted to charge directly. In this way, no intermediate step in producing light is required. The electrons and holes are directed by the electric field towards the TFT elements, thus eliminating the lateral spreading of the signals during transit, resulting in high spatial resolution images.

4.3 PHOTON YIELD

Problem:

A 60 keV X-ray photon strikes the phosphor of a CR plate, producing light photons with a wavelength of 415 nm. The energy conversion coefficient for this process is 18%. How many light photons are produced?

Solution:

Calculate the energy for the X-ray photon,

$$\varepsilon_{60\,keV} = 60\times10^3\times1.602\times10^{-19}\,J = 9.612\times10^{-15}\,J$$

Only 18% of the X-ray photons are converted to light,

$$\varepsilon_{x-ray} = 9.612\times10^{-15}\,J\times0.18 = 1.73\times10^{-15}\,J$$

Calculate the energy quanta for a 415 nm light photon,

$$\varepsilon_{light} = \frac{h\nu}{\lambda}$$

where,

h is Planck's constant $= 6.6\times10^{-34}\,Js$

v is the speed of light $= 3 \times 10^8 \, \text{ms}^{-1}$

λ is the wavelength of the photon

$$\varepsilon_{\text{light}} = \frac{hv}{\lambda} = \frac{6.6 \times 10^{-34} \, \text{Js} \times 3 \times 10^8 \, \text{ms}^{-1}}{415 \times 10^{-9} \, \text{m}} = 4.77 \times 10^{-19} \, \text{J}$$

Thus, the number of light photons produced, N

$$N = \frac{1.73 \times 10^{-15} \, \text{J}}{4.77 \times 10^{-19} \, \text{J}} = 3626$$

4.4 IMAGE STORAGE

Problem:

a. Calculate the required storage capacity (in megabytes, MB) to store an uncompressed 10 seconds, 8-bit grayscale ultrasound video file of 512 × 512 pixel dimensions, given that the frame rate for the video is 30 frames/seconds.

b. What would be the new required storage if the video were in RGB colour format instead?

c. A computed radiography (CR) image with an image matrix of 2048 × 2500 was used to image a field of view of 35.56 × 43.18 cm². What is the pixel size? The CR image is stored with 10 bit/pixel. What is the bit depth for this image?

Solution:

a. Storage capacity for 1 frame = (512 × 512) pixel × 8 bit/pixel = 2,097,152 bit

For 30 frames and 10 seconds,

Storage capacity = 2,097,152 bit/frame × 30 frame/sec × 10 sec

$= 629{,}145{,}600$ bit

$= 78{,}643{,}200$ byte

$\approx 75 \, \text{MB}$

b. To store the video in RGB colour format, the new storage capacity needed would be

$78{,}643{,}200 \times 3 = 235{,}929{,}600$ bytes $\approx 225 \, \text{MB}$

c.
$$\text{Pixel size} = \left(\frac{35.56 \text{ cm}}{2048 \text{ pixels}} \right) \times \left(\frac{43.18 \text{ cm}}{2500 \text{ pixels}} \right)$$

$$= 0.017 \times 0.017 \text{ cm}^2$$

Bit depth $= 2^{10} = 1024$

Note:

1 byte = 8 bits
1 kB = 1024 bytes
1 MB = 1024^2 bytes

4.5 GREY SCALE

Problem:

What is a grey-scale histogram? Sketch the histograms for the two radiographs shown in Figure 4.3.

Describe in what situation we would apply histogram equalization.

Solution:

Grey-scale histogram is a function showing, for each grey level, the number of pixels in the image that have that grey level. Figure 4.4 shows the grey scale histogram of the two radiographs. We apply histogram equalization so that the grey-levels will take advantage of the full window width of the display.

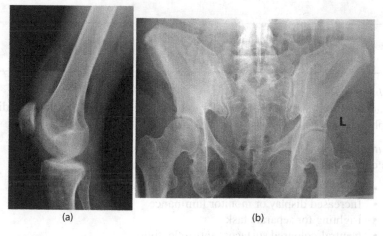

(a)   (b)

FIGURE 4.3 Two radiographs showing (a) knee and (b) pelvis region.

(a) Frequency

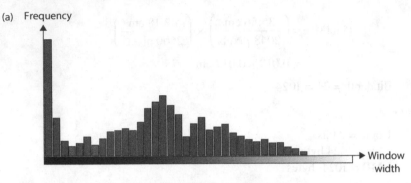

(b) Frequency

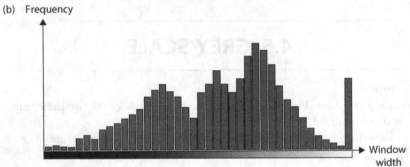

FIGURE 4.4 Grey-scale histogram of the radiographs (a) knee and (b) pelvis region.

4.6 SOFT COPY REPORTING

Problem:
What conditions are required in order to optimise softcopy reporting of radiographic images?

Solution:
The conditions are:

- Decreased room luminance (ambient light level)
- Increased display or monitor luminance
- Lighting for separate task
- Neutral-coloured surfaces, anti-reflectivity
- Partitions to separate individual workstations

Note:

Practice guidelines on softcopy reporting according to American College of Radiology (ACR) – American Association of Physicists in Medicine (AAPM) – Society of Imaging Informatics in Medicine (SIIM) (Norweck et al., 2013).

- Recommended ambient lighting level: 20–40 lux.
- The display luminance for diagnostic monitors should be at least 350 cd m^{-2} and for mammography should be 420 cd m^{-2}.

Note: 1 lux = 1 cd.sr m^{-2}

4.7 GREY SCALE

Problem:

Define and explain briefly the concept of 'fill factor' as used in flat panel detectors.

Solution:

Fill factor is defined as $\dfrac{\text{light sensitive area}}{\text{area of detector element}}$

Because the electronics on a detector element are fixed, the efficiency depends on the ratio of the light sensitive area to the physical size of the detector element.

4.8 COMPUTED RADIOGRAPHY

Problem:

a. Describe the detector used in computed radiography (CR).
b. Explain the process of photostimulable luminescence in computed radiography.

Solution:

a. The detector in CR is a photostimulable phosphor (PSP) (also known as storage phosphor). It is typically made of powdered barium fluorohalide, which contains a small amount of europium as a doping agent to increase its sensitivity. This powder is mixed with resin and laid down as a ~0.3 mm thick coating onto a conductive layer, which in turn is laid on a support layer and laminated base (Figure 4.5).

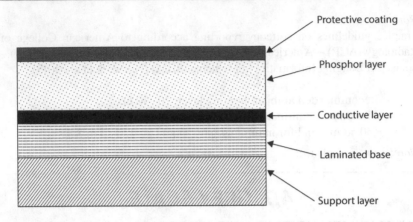

FIGURE 4.5   Structure of PSP.

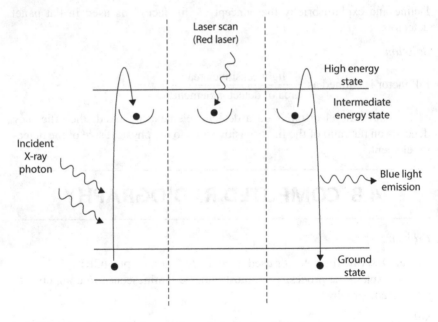

FIGURE 4.6   Principle of operation of photostimulable luminescence.

    b. Computed radiography is based on the physical process of photo-stimulable luminescence (Figure 4.6).

Excitation – When exposed to X-rays, electrons are excited from a lower energy state to a higher energy state where they are trapped.

Latent image – Latent image is stored in the intermediate energy state. This is created by the dopant that makes the excited electrons state for a workable lifetime.

Luminance – The 'read-out' by stimulating the trapped electrons using red/infrared energy; as the electrons are moved to the lower energy states, they emit visible light in the blue region.

Note:

The other PSP material available is caesium bromide doped with europium ($CsBr:Eu^{2+}$), which can be manufactured in a 'structured' format, enabling higher spatial resolution and detection.

4.9 EXPOSURE INDEX AND DEVIATION INDEX

Problem:

Explain the term exposure index (EI) and deviation index (DI) used in digital radiography.

Solution:

Exposure index (EI) indicates the amount of radiation exposure to the digital detector. It does not indicate the amount of dose received by the patient. Deviation index (DI) is an index that was published by the International Electrotechnical Commission (IEC) IEC 62424-1. The DI takes into account the target exposure index (EI_{target}) and compares the exposure.

$$DI = 10\log_{10}\left(\frac{EI}{EI_{target}}\right)$$

Note:

To date, all DR manufacturers have a different EI system, see example in Table 4.1.

TABLE 4.1 Exposure indicator by different manufacturer

MANUFACTURER	EXPOSURE INDICATOR NAME	SYMBOL	UNITS
Carestream	Exposure index	EI	Mbels
GE	Uncompensated detector exposure	UDExp	µGy air kerma
GE	Compensated detector exposure	CDExp	µGy air kerma
GE	Detector exposure index	DEI	Unitless
Fuji	S value	S	Unitless
Agfa	Log of median of histogram	lgM	Bels
Siemens	Exposure index	EXI	µGy air kerma
Philips	Exposure index	EI	Unitless

Source: Adapted from Seibert, J. A. and Morin, R. L., *Pediatric Radiology*, 41, 573–581, 2011.

Image Quality

5

5.1 QUALITY ASSURANCE

Problem:
Discuss the goal of quality assurance (QA) in radiology.

    a. List and define the three physical parameters used to describe image quality of a general radiograph. Discuss how these three parameters affect each other.

    b. Explain why quality control (QC) on focal spot size is important. Discuss briefly the QC test for this purpose.

Solution:
QA is to maintain optimal diagnostic image quality with minimum risk and distress to patient and also to be cost effective.

    a. Three physical parameters that are used to describe image quality of a general radiograph are contrast, noise and spatial resolution.

       Contrast is the difference in the grey scale between closely adjacent regions on the image.
       Noise is the uncertainty in a signal due to random fluctuations in that signal.
       Spatial resolution is the ability to image and visually distinguish two separate objects.

Increase in image noise will result in a decreased image contrast. The overall detectability of details increase with higher signal-to-noise ratio.

    b. Focal spot size determines geometric blur, i.e. spatial resolution. As the X-ray tube ages, the anode target becomes pitted, which broadens the focal spot and degrades image resolution. Focal-spot size should be evaluated annually or whenever an X-ray tube is replaced. The pinhole camera is difficult to use and requires excessive exposure time. The star pattern is easy to use but has significant limitations for focal-spot sizes less than 0.3 mm. The standard for measurement of effective focal-spot size is the slit camera. An acceptable alternative to focal-spot size measurement is use of a line-pair test tool to determine limiting spatial frequency. Specification of focal-spot size depends not only on the geometry of the tube but also on the focusing of the electron beam. Consequently, vendors are permitted a substantial variance from their advertised focal-spot sizes.

5.2 CONTRAST

Problem:
What is contrast? Differentiate subject contrast, detector contrast and displayed contrast.

Solution:
Subject contrast is the difference in the X-ray beam intensities due to attenuation differences in patient. This is before the X-ray photons are incident

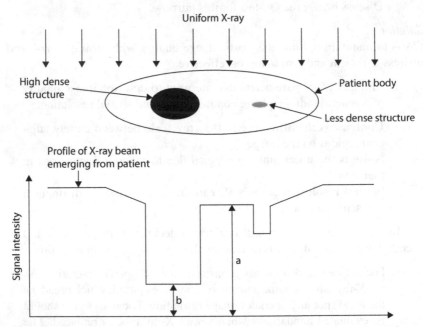

FIGURE 5.1   X-ray beam incident upon a patient. The X-ray photons are attenuated by different parts of the body resulted in non-homogeneous output X-ray photons hitting the detector.

upon the detector. It is influenced by both the intrinsic and extrinsic factors. Intrinsic factors are the actual anatomical (thicker or thinner body parts) or functional change in patient tissue (lesion denser than normal tissue), which will be depicted as brighter or darker objects on radiographs. Extrinsic factors are the adjustment of image acquisition protocols that are used to enhance object contrast (adjust kVp, mAs, and the use of contrast medium).

Subject contrast is denoted as, $C_s = \dfrac{S_a - S_b}{S_a}$, $S_a > S_b$ (refer to Figure 5.1).

Receptor contrast is the contrast that results when the incident photon beam reaches the detector/receptor. Receptor contrast can change the subject contrast and it is influenced by the characteristic curve of the receptor. For example, for a screen film radiograph, the characteristic curve is a sigmoidal curve. The linear slope is called gamma, γ.

$$\gamma = \frac{D_2 - D_1}{\log E_2 - \log E_1}$$

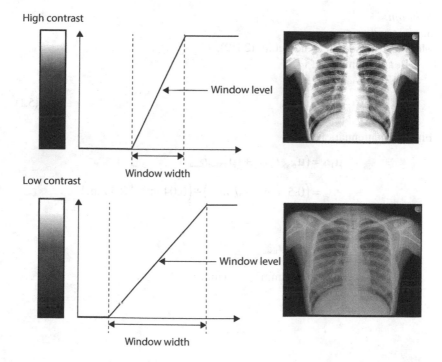

FIGURE 5.2 Displayed contrast.

A high gamma is reflected as a steep curve and narrow latitude. This will result in an image with high contrast.

For a digital detector, the contrast is depicted as a linear response with input signal due to the linear dynamic range.

Displayed contrast refers to the adjustable contrast obtained from a digital image. It displays all the grey levels in a limited number of pixel levels. It uses a lookup table, window level and window width to achieve the desired contrast level (Figure 5.2).

Problem:
A phantom is constructed with a piece of bone embedded in soft tissue-like material. Ignoring the effect of scatter, calculate the subject contrast between the X_1 and X_2 and the radiographic contrast (Figure 5.3).
Given,

> Film gamma = 3
> Linear attenuation coefficient of bone = 0.5 mm^{-1}
> Linear attenuation coefficient tissue = 0.04 mm^{-1}

Solution:
Subject contrast:
Method 1: Based on Dendy et al. (2012),
  Beam attenuation,

$$X_1 = X_0 e^{-\mu_1 t_1}$$

$$X_2 = X_0 e^{-\mu_2 t_2}$$

(5.1)

Find total attenuation,

$$\mu_1 t_1 = \left(\mu_{\text{bone}} t_{\text{bone}}\right) + \left(\mu_{\text{tissue}} t_{\text{tissue}}\right)$$

$$= \left(0.5 \text{ mm}^{-1} \times 7 \text{ mm}\right) + \left(0.04 \text{ mm}^{-1} \times 3 \text{ mm}\right)$$

$$= 3.62$$

$$\mu_2 t_2 = \mu_{\text{tissue}} t_{\text{tissue}}$$

$$= 0.04 \text{ mm}^{-1} \times 10 \text{ mm} = 0.4$$

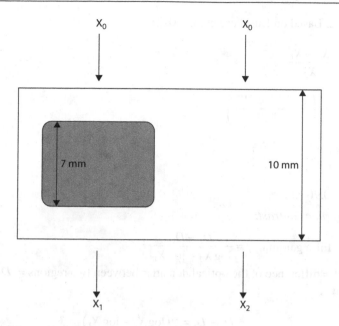

FIGURE 5.3 Subject contrast.

Subject Contrast,

$$C_{\text{subject}} = \log_{10} \frac{X_2}{X_1} \tag{5.2}$$

By applying the logarithmic identity $\log_{b^a} = \dfrac{\log_{c^a}}{\log_{c^b}}$ to change the base, we have an expression in natural logarithm,

$$C_{\text{subject}} = \frac{\ln \dfrac{X_2}{X_1}}{\ln 10} = \frac{1}{\ln 10} \left(\ln \frac{X_2}{X_1} \right) = 0.434 \left(\ln X_2 - \ln X_1 \right) \tag{5.3}$$

Substitute Equation (5.1) in Equation (5.3), assume $X_o = 1$,

$$
\begin{aligned}
C &= 0.434 \left[\left(\ln X_o e^{-\mu_2 t_2} \right) - \left(\ln X_o e^{-\mu_1 t_1} \right) \right] \\
&= 0.434 \left[\left(\mu_1 t_1 \right) - \left(\mu_2 t_2 \right) \right] \\
&= 0.434 \left[3.62 - 0.4 \right] \\
&= 1.4
\end{aligned}
\tag{5.4}
$$

Method 2: Based on Bushberg et al. (2012),

$$C = \frac{X_2 - X_1}{X_2}$$

$$= X_0 \left(\frac{e^{-\mu_2 x_2} - e^{-\mu_1 x_1}}{e^{-\mu_2 x_2}} \right) \tag{5.5}$$

$$= \left(\frac{e^{-0.4} - e^{-3.62}}{e^{-0.4}} \right)$$

$$= 0.96$$

Radiographic contrast:

$$\text{Film gamma, } \gamma = \frac{D_2 - D_1}{\log X_2 - \log X_1} \tag{5.6}$$

Contrast = difference of the optical densities between two regions = $D_2 - D_1$
Rearrange,

$$D_2 - D_1 = \gamma \left(\log X_2 - \log X_1 \right)$$

Substitute Equation (5.3),

$$D_2 - D_1 = \gamma \cdot 0.434 \left(\mu_1 x_1 - \mu_2 x_2 \right)$$

$$= 3 \times 1.4$$

$$= 4.2$$

5.3 LINE SPREAD FUNCTION

Problem:
The line spread functions (LSFs) in Figure 5.4 are produced by two different phosphor layers of a CR, one thicker, the other thinner. Indicate which LSF profile is produced by each of the phosphor layers and explain the reason.

Solution:
LSF(A) is produced by a thicker phosphor layer while LSF(B) is produced by a thinner phosphor layer. Assuming that the detection efficiency of the phosphors was the same, the detection probability of the X-ray photons would be

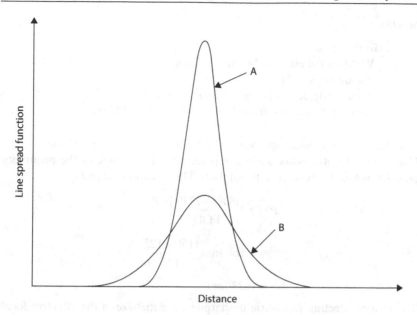

FIGURE 5.4   Line spread profiles from two different phosphor layers.

similar for both phosphors. Therefore, the thicker phosphor will produce more light photons. As the number of photons increases, the line spread function, which actually represents the Poisson distribution of the photons, will become narrower.

5.4 GEOMETRIC PENUMBRA

Problem:

a. The width of the electron beam in an X-ray tube is 5.5 mm, and the anode angle is 12°. A patient is laid in a supine position at 150 cm from the X-ray tube, directly on top of the flat panel. The patient has a small lesion located near the surface of his abdomen. Calculate the geometric unsharpness produced by the lesion, given that the tissue thickness is 25 cm.

b. Discuss the factors influencing the geometric unsharpness of an image, and methods to minimise this effect.

Solution:

a. Given,
   Width of the electron beam = 5.5 mm,
   Anode angle = 12°,
   Focus to detector-distance (FDD) = 150 cm
   Focus to object-distance (FOD) = (150–25) = 125 cm

The effective X-ray focus spot size, $f = 5.5$ mm $\times \tan 12° = 1.169$ mm
Geometric unsharpness of an image is reflected by the size of the penumbra (also known as the geometric penumbra). This is expressed as P,

$$P = f \frac{(\text{FDD} - \text{FOD})}{\text{FOD}}$$

$$= 1.169 \text{ mm} \frac{(150 - 125)\text{cm}}{125 \text{ cm}}$$

$$= 0.23 \text{ mm}$$

The factors affecting geometric unsharpness are the size of the effective focal spot of the X-ray tube, the FOD, FDD and the object imaged. To improve the spatial resolution of the image, we need to either have a small focus spot, and/or increase the FOD.

5.5 CONTRAST WITH RELATIVE SCATTER

Problem:
The amount of scattered radiation emerging from a scattering body, such as a patient, varies with the X-ray tube potential. When it impinges on the image receptor, scattered radiation will result in reduced image contrast. Discuss the measures that can be taken to reduce the loss of contrast due to scattered radiation.

Solution:
Methods to minimise scattered radiation:

- Apply collimation to reduce size of the X-ray beam to cover area of interest only. This reduces the tissue volume exposed to radiation.
- The use of lower kVp will reduce scattered radiation reaching the detector. However, this technique is limited by the X-ray penetration

required to achieve image with diagnostic qualities. Also, it will increase the radiation dose received by the patient due to the need to increase the tube current-time (mAs).

- Adjust the orientation of the patient such that the region of interest is closer to the receptor. This will reduce scattered radiation due to geometric effect. For example, if the ROI is the left ear, left ear is placed closer to receptor.
- Apply compression to the body parts that required imaging. In this way, the soft tissues are forced out of the primary beam therefore reducing scattering volume.
- Use of anti-scattered grids.
- Apply air gap technique. This technique is used in mammography magnification view. Scattered radiation is comprised of many photons traveling in trajectories taking them away from the image receptor. The presence of air gap between the imaged object and receptor may result in most of the scattered radiation being deflected out and not reaching the receptor thus increasing image contrast.

Note:

Figure 5.5 shows the amount of scattered radiation as a fraction of the primary radiation as a function of tube potential.

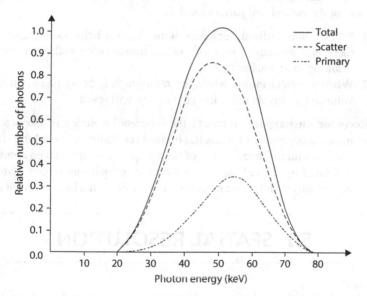

FIGURE 5.5 Amount of scattered radiation as a fraction of the primary radiation as a function of tube potential. (From Dendy, P. P. and Heaton, B., *Physics for Diagnostic Radiology*, CRC Press, Boca Raton, 2012.)

5.6 RESOLUTION

Problem:
Assuming a radiographer had applied all the necessary techniques to obtain a high quality radiograph, discuss the factors that could affect the resolution of a radiograph that are beyond the control of a radiographer.

Solution:
Factors that could affect the resolution of a radiograph that are beyond the control of the radiographer are the geometric unsharpness and unsharpness due to patient and receptor.

Geometric unsharpness refers to the finite size of the X-ray tube focal spot. Since the focal spot cannot be infinitesimally small, it will inadvertently produce a broader penumbra, called geometric penumbra. The size of the focal spot is limited by the heat rating considerations; hence, it is not possible to reduce it to be infinitely small. As a result, very small objects such as microcalcifications may not be visualised if the geometric penumbra is too large. Increasing the focus to image distance can reduce the size of the geometric penumbra.

Patient unsharpness may be caused by two factors, the non-uniform thickness of the patient and patient motion.

1. When a non-uniform thickness of an object is being radiographed, there is a gradual change of photon transmission and absorption resulting in an unsharp edge.
2. When a patient moves while the radiograph is being taken, either voluntary or involuntary, image blurring will result.

Receptor unsharpness refers to the inherent resolution of the detector. For example, radiographic film has the highest resolution, whereby the size of the grain in the film is the resolution of the receptor. For film-screen combination, it is limited by the light spread within the intensifying screen, while in digital radiography systems, the detector element size is the limiting resolution.

5.7 SPATIAL RESOLUTION

Problem:
Spatial resolution is one of the factors affecting the image quality. Modulation transfer function (MTF) is a graphical description of the resolution characteristics of an imaging system. In digital imaging, it is what defines or limits the spatial resolution

of the imaging device. However, despite this limitation, why is the MTF of most digital imaging systems still superior to those of the screen film technology?

Solution:
Detector element size determines the spatial resolution of an imaging system. However, with advanced image post-processing techniques, the MTF of most digital imaging systems is still superior to those of the screen film technology. Digital imaging systems has a wider dynamic range. For example, a digital radiography system has a bit depth of 2^{14} (equal to 16384 shades of gray). Postprocessing techniques (e.g. adjusting the window level and width) enable better visualization of all shades of gray. In contrast, a screen film system has only essentially 3 orders of magnitude in the optical density changes (equal to 1000 shades of gray), and once the film is develop, nothing can be done to change the different shades of gray. Digital detectors also have higher detective quantum efficiency (DQE), which also improves MTF.

Problem:
Define spatial resolution in medical imaging and what is meant by a line pair.

    a. A digital radiographic imaging system has a spatial resolution of 2.5 lp mm^{-1}. How small an object can it resolve?

    b. What is the maximum resolvable spatial frequency of a CT image with a matrix of 512×512 and a field of view of 25 cm?

Solution:
Spatial resolution is the ability of an imaging system to discriminate between two adjacent high-contrast objects. Line pair describes a high-contrast line separated by an interspace of equal width, which is often used to quantify the spatial resolution of an imaging system or modality.

    a. For a system with a spatial resolution of 2.5 lp mm^{-1}, five objects can be distinguished within 1 mm. Therefore, the smallest object resolvable with this system would be 1/5 mm or 0.2 mm.

    b. $\text{Pixel size} = \dfrac{25 \text{ cm}}{512 \text{ pixels}} = 0.049 \text{ cm}$

Two pixels are required to form a line pair; therefore:

$$2 \times 0.049 \text{ cm} = 0.098 \text{ cm lp}^{-1}$$

$$\frac{1 \text{ lp}}{0.098 \text{ cm}} = 10.2 \text{ lp cm}^{-1}$$

5.8 MODULATION TRANSFER FUNCTION

Problem:

a. Explain modulation transfer function (MTF).
b. How is the modulation transfer function (MTF) affected when a system increases magnification?
c. In an imaging chain, if PART A and PART B of an imaging system have MTFs of 0.6 and 0.8, respectively, at a certain spatial resolution, what is the overall MTF of the system?
d. Figure 5.6 shows the plot of MTF against spatial frequency. Compare the spatial resolution of systems *A, B* and *C*.

Solution:

a. MTF describes the capacity of an imaging detector to transfer the modulation of the input signal at a given spatial frequency to its output. It is expressed as the ratio of the output and input signal modulation.
b. When the magnification of the system is increased, imaging of details improves; hence, the spatial resolution improves. However,

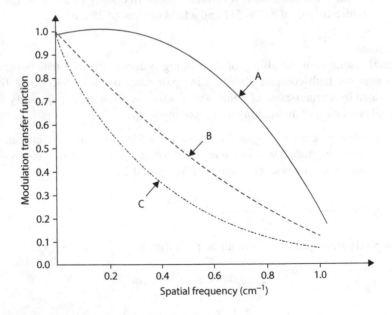

FIGURE 5.6 Modulation transfer function.

the magnification of the system may also increase geometric unsharpness thus compromising the improvement in the MTF.

c. The overall MTF of the imaging system is the product of the MTF of its components.

$$MTF_{overall} = MTF_1 \times MTF_2 = 0.6 \times 0.8 = 0.48$$

d. System *A* has the highest spatial resolution, followed by systems *B* and *C*.

Note: For further readings, please refer to Dance et al., 2014.

Problem:

The following 3 × 3 matrices (*a*) to (*d*) are filter kernels. Plot the MTF of the filtered image on a single MTF graph.

(a)			(b)			(c)			(d)		
0	0	0	1	1	1	1	1	1	1	1	1
0	1	0	1	1	1	1	5	1	1	10	1
0	0	0	1	1	1	1	1	1	1	1	1

Solution:

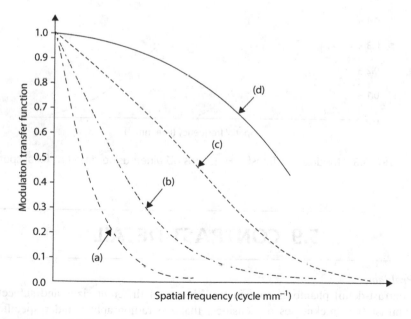

FIGURE 5.7 MTF resulting from different filter kernels.

Problem:

The MTFs in Figure 5.8 are produced by two different X-ray tubes, one with a focal spot of 1 mm, the other with a focal spot size of 5 mm. Match each of the MTF with the correct focal spot size and explain the reason.

Solution:

MTF(*A*) is produced by the X-ray tube with focal spot size of 1 mm while MTF(*B*) is produced by the X-ray tube with focal spot size of 5 mm. A smaller focal spot size will result in higher spatial resolution of the image, which is reflected as a slower fall off of the MTF curve.

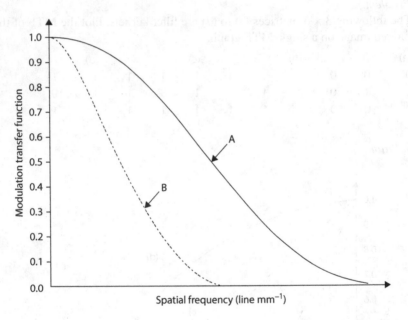

FIGURE 5.8   Modulation Transfer Functions obtained using different focal spot sizes.

5.9 CONTRAST-DETAIL

Problem:

Contrast-detail phantoms contain test objects of different sizes and subject contrast (i.e. thicknesses or densities) that are radiographed under specific exposure conditions.

a. How does noise affect image contrast?
b. Describe the theoretical basis for contrast-detail phantoms that produce the contrast-detail curves as shown in Figure 5.9.
c. Comment on the graph shown in Figure 5.9 obtained from an experiment on a mammography system. What conclusions could you make?

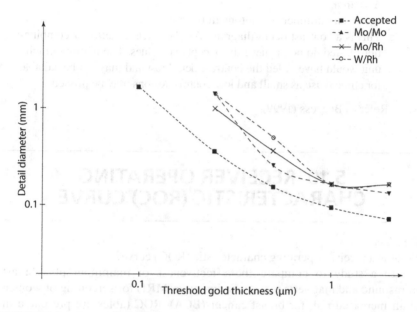

FIGURE 5.9 Contrast-detail diagram.

Solution:

a. Higher noise reduces the signal-to-noise ratio (SNR) or the ability to distinguish between the differences in intensity in an image and therefore reduces image contrast.
b. The theoretical basis of the Rose Model: The visibility of an object in a background is a complex phenomenon involving technical factors as well as subjective factors, such as visual perception. The Rose Model is based on SNR, and evaluates the quality of digital radiographic images. The Rose Model is utilised to assess the detectability of signals by human observation, providing a description of an object's visibility in an image.

Contrast threshold of C_T is given by

$$C_T = \frac{k}{\sqrt{A(n_b)}}$$

where,

    k = threshold SNR,

    A = area,

    n_b = mean number of photons in the area

  c. This is a contrast-detail diagram. All the three target/filter combinations tested do not achieve the acceptable values. The mammography unit would have failed the contrast-detail test and may not be suitable for clinical use as small and low contrast lesions may be missed.

Note: Refer to Burgess (1999).

5.10 RECEIVER OPERATING CHARACTERISTIC (ROC) CURVE

Problem:

What is a receiver operating characteristic (ROC) curve?

In a study to compare the effectiveness of mammography, breast ultrasound and magnetic resonance imaging (MRI) for screening of women at an increased risk for breast cancer (BCA), ROC tables are presented in Table 5.1.

  a. Determine the (i) sensitivity, (ii) specificity, (iii) accuracy, (iv) positive predictive value, and (v) negative predictive value of each system.

  b. Which modality would you recommend for breast cancer screening?

  c. If a test is always positive, what is its sensitivity and specificity?

Solution:

  a. The better the image quality, the higher the accuracy of diagnosis. ROC is a useful tool to evaluate image quality of radiological images. It is useful when considering a two-class prediction problem, in which the outcome is either positive or negative (e.g. either the patient has BCA or is normal). The ROC can only be used if the ground truth is known (i.e. from biopsy, etc.). The radiologist, upon evaluating the radiological images, needs to make a decision if he thinks the patient has BCA or is normal. Hence, there are four possible outcomes, shown in Table 5.2.

The ROC is used to report the following measures,

- Sensitivity or true positive rate (TPR)

$$\text{Sensivitity} = \frac{\text{number of patients correctly diagnosed as having BCA}}{\text{actual number of patients having BCA}}$$

$$= \frac{TP}{TP + FN}$$

TABLE 5.1 ROC tables comparing the effectiveness of (a) mammography, (b) breast ultrasound and (c) MRI in detecting breast cancer

(A) MAMMOGRAPHY			(B) BREAST ULTRASOUND			(C) MRI		
ACTUAL CONDITION			*ACTUAL CONDITION*			*ACTUAL CONDITION*		
Diagnosis	*BCA*	*Normal*	*Diagnosis*	*BCA*	*Normal*	*Diagnosis*	*BCA*	*Normal*
BCA	3	7	*BCA*	3	19	*BCA*	9	5
Normal	6	89	*Normal*	6	77	*Normal*	0	91

TABLE 5.2 ROC table and the four possible outcomes

	ACTUAL CONDITION	
Diagnosis	*BCA*	*Normal*
BCA	True positive (TP)	False positive (FP)
Normal	False negative (FN)	True negative (TN)

- Specificity or true negative rate (TNR)

$$\text{Specificity} = \frac{\text{number of patients correctly diagnosed as healthy}}{\text{actual number of healthy patients}}$$

$$= \frac{TN}{TN + FP}$$

- False positive rate (FPR)

$$FPR = \frac{\text{number of healthy patients wrongly diagnosed as having BCA}}{\text{actual number of healthy patients}}$$

$$= \frac{FP}{TN + FP}$$

- Accuracy

$$Accuracy = \frac{\text{number of correct diagnosis}}{\text{total number of patients}} = \frac{TP + TN}{TP + TN + FP + FN}$$

- Positive predictive value (PPV) or precision

$$PPV = \frac{\begin{array}{c}\text{number of patients correctly} \\ \text{diagnosed as having BCA}\end{array}}{\text{total number of patients having BCA}} = \frac{TP}{TP + FP}$$

- Negative predictive value (NPV)

$$NPV = \frac{\text{number of patients correctly diagnosed as healthy}}{\text{total number of patients diagnosed as healthy}} = \frac{TN}{TN + FN}$$

b. Based on the sensitivity of the systems (Table 5.3), MRI would be the best system to detect breast cancer. However, it may not be a viable modality for mass BCA screening due to high cost, accessibility and time consumed to perform the imaging.

c. If a test is always positive, the sensitivity is 1 (100%) and specificity is 0.

TABLE 5.3 Effectiveness of mammography, breast ultrasound and MRI for screening of women at an increased risk for BCA

	SENSITIVITY	SPECIFICITY	ACCURACY	PPV	NPV
Mammography	33%	93%	88%	30%	94%
Breast US	33%	80%	76%	14%	93%
MRI	100%	95%	95%	64%	100%

Problem:
What is the function of an ROC curve (Figure 5.10) for a specific imaging procedure? How is the human operating point in the figure below being determined? Where are an ideal test and a no-predictive-value test to be located on the curve?

Solution:
ROC curve measures the diagnostic accuracy of a medical imaging system and illustrates the performance of a binary classifier system.

The human operating point is chosen on the curve that is closest to the top left corner. An ideal test would have 100% sensitivity and 100% specificity, which is on the top left corner. No-predictive value would be along the diagonal. Right bottom corner would be 100% predictive of the opposite of truth.

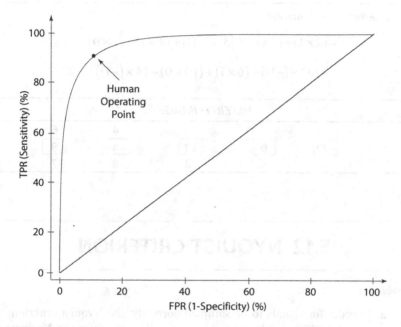

FIGURE 5.10 ROC curve.

5.11 FILTER KERNEL

Problem:
The following image matrix is applied with a filter. Calculate the pixel value in *A, B, C* and *D*.

ORIGINAL IMAGE						FILTER				FILTERED IMAGE					
2	5	3	5	4	6	1	0	−1		2	5	3	5	4	6
4	3	32	5	6	9	1	0	−1	=	4	A	B	C	D	9
6	10	4	8	8	7	1	0	−1		6	10	4	8	8	7

(* between FILTER columns indicates convolution)

Solution:

A = −27			B = 0			C = 21			D = −4		
2	0	−3	5	0	−5	3	0	−4	5	0	−6
4	0	−32	3	0	−5	32	0	−6	5	0	−9
6	0	−4	10	0	−8	4	0	−8	8	0	−7

Example of calculations:

$$A = (2 \times 1) + (5 \times 0) + (3 \times (-1)) + (4 \times 1) + (3 \times 0)$$

$$+ (32 \times (-1)) + (6 \times 1) + (10 \times 0) + (4 \times (-1)) = -27$$

		FILTERED IMAGE			
2	5	3	5	4	6
4	**−27**	**0**	**21**	**−4**	**9**
6	10	4	8	8	7

5.12 NYQUIST CRITERION

Problem:

a. In order for signals to be sampled correctly, the Nyquist criterion must be fulfilled; otherwise, aliasing will occur. State the Nyquist criterion and by means of a diagram, explain briefly the aliasing effect. Give two examples of aliasing effect observed in medical imaging applications.

b. You are given a 36 cm × 43 cm chest radiograph to be digitised. The requirement is 16 bits per pixel, with a pixel spacing that preserves the intrinsic spatial resolution in the chest radiograph (5 lp mm^{-1}). Calculate the storage memory needed.

Solution:

a. **Nyquist criterion** – the signal must be sampled at least twice in every cycle or period, i.e. the sampling frequency must be at least twice the highest frequency present in the signal. The maximum signal frequency that can be accurately sampled is called the Nyquist frequency. For example, the Nyquist frequency for a system that is sampled at a frequency of 100 Hz is 50 Hz.

Aliasing occurs when high frequency signals are erroneously recorded as low frequency signals. High frequency components carrying information about small structures with sharp edges will be lost.

Some examples:

- In MRI, aliasing causes the appearance of wrap-around, translocation of anatomy from one side of the image to the other.
- In CT, the body section is sampled by a fan beam consisting of a limited number of X-ray pencil beams, coming from a limited number of directions. If these are too few, the system does not have the information to reproduce sharp high-contrast boundaries, instead 'low frequency' streak artefacts are produced.
- In pulsed Doppler ultrasound, the flow of blood cells is sampled by a series of ultrasound pulses. If the pulse repetition rate is not fast enough, fast flow in one direction will be interpreted as a slower flow in the opposite direction.

b. To preserve 5 lp mm^{-1} resolution of the chest radiograph, we need 10 samples per mm or 0.1 mm per pixel.

$$\text{Number of pixels} = 3600 \times 4300$$

$$= 3600 \times 4300 \times 2 \text{ bytes}$$

$$= 30,960,000 \text{ bytes}$$

$$= 30.96 \text{ Mbytes}$$

Note:

In medical imaging, we often use lp mm^{-1} instead of cycle mm^{-1} when referring to an analogue image.

Mammography

6

6.1 TARGET-FILTER

Problem:

Sketch the output spectra from a mammography system operating with 30 kVp tube voltage when the following target-filter combination is installed: Mo-Mo, Mo-Rh, Rh-Rh, W-Rh, Rh-Al, Mo-Al, and Rh-Mo. Comment on the use of rhodium and tungsten as target material for mammography.

Solution:

The output spectra for different target-filter combinations are shown in Figures 6.1 through 6.3.

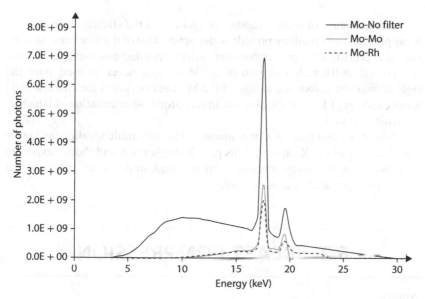

FIGURE 6.1 Output spectra for Mo anode tube, without filter and with different filter combinations.

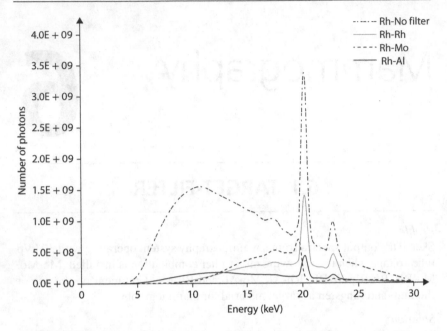

FIGURE 6.2 Output spectra for Rh anode tube without filter and with different filter combinations.

The advantage of using a rhodium target is that the slightly higher energy X-ray photons from rhodium provide better penetration of thick or dense breasts. This also decreases the patient dose but suffers from decreased image contrast.

The target-filter combination of Rh/Mo should never be used since the high attenuation beyond the k-edge of the Mo filter occurs at the characteristic X-ray energies of Rh. This will result in inappropriate attenuation of Rh characteristic X-rays.

Tungsten target produces large amount of bremsstrahlung, allowing higher power loading of the X-ray unit. This permits higher mA and shorter exposure time. However, the X-ray spectrum contains undesirable L-edge X-rays that need to be attenuated to an acceptable level.

6.2 BREAST COMPRESSION

Problem:
During a mammography examination, the breast being examined is compressed with a paddle. Explain the reasons for breast compression.

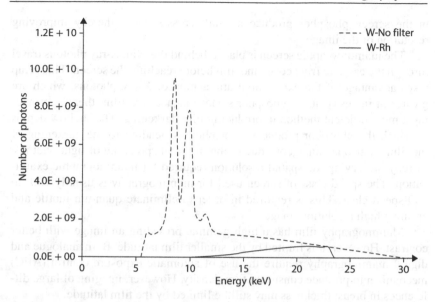

FIGURE 6.3 Output spectra for tungsten anode.

Solution:
Breast compression is needed to make the breast thickness uniform across the image field-of-view so that the tissue can be accurately visualised. Spreading out the tissue also ensures that small lesions will not be obscured by overlying breast tissue. Compression helps to keep the breast still during the examination, eliminating image blurring due to motion. A compressed breast is thinner, reducing scattered radiation to the image receptor and increasing the sharpness of the image. The required X-ray dose for a given image signal-to-noise ratio is lower with compression; this subsequently reduces the mean glandular dose of the breast.

6.3 SCREEN FILM AND DIGITAL RECEPTOR

Problem:
Compare and contrast film-screen mammography and digital mammography.

Solution:
Film-screen mammography uses single emulsion-single screen or double emulsion-double screen combination in order to eliminate parallax error. Rare earth screens, such as gadolinium oxysulphide: terbium-activated (Gd_2O_2S:Tb), are commonly used in mammography. Needle-shaped fibre optic-like crystals

in the screen phosphor produce a small cross-section, thereby improving resolution of the image.

The mammographic screen is placed behind the film. X-ray photons travel through the cassette front cover and film before reaching the screen. This setup takes advantage of the exponential attenuation of X-ray photons, which are greater in intensity at the phosphor surface adjacent to film, thereby providing a more efficient method in producing film darkening. The shallow depths in which the phosphor produces light photons, leading to the darkening of the film, is also advantageous due to the minimal spreading of light photons, thereby preserving the spatial resolution required for mammographic examination. The speed class of screen used for mammography is usually 100- to 200-speed class. This is required in order to eliminate quantum mottle and ensure a high resolution image.

Mammography film has a high gamma, providing an image with better contrast. However, it is limited by the smaller film latitude. Both analogue and digital mammography require the use of automatic exposure control (AEC) mechanism to produce consistent film density. However, imaging of large differences in breast thickness may still be limited by the film latitude.

Digital mammography uses digital receptors. The advantage of digital receptors is due to its larger exposure dynamic range (1000:1), which can overcome the exposure latitude limitation of screen-film mammography, providing better image quality at a lower radiation dose. A drawback of the digital receptors is the fact that the intrinsic resolution of the digital receptors is inferior to screen film mammography. This is because for screen film, the grain size of the film is the intrinsic resolution, while in digital receptors it is limited by the pixel/detector size. However, the post-processing image, combined with the large dynamic range of exposure, can improve the image quality. A typical matrix size in digital mammography is 4 × 6 k, with a pixel size of 40–50 μm. The limiting spatial resolution of digital mammography is about 10 lp/mm. This is inferior to screen/film, which can reach 15 lp/mm.

6.4 MAMMOGRAPHY X-RAY TUBE

Problem:
What are the special requirements of mammography imaging and the demand on the X-ray tube? How are these achieved with a

    a. Molybdenum anode tube?
    b. Tungsten anode tube?

Solution:

Mammography is the imaging of breast is comprised primarily of fat and fibroglandular tissues. The atomic densities of these tissues are similar, while their effective atomic numbers (Z) differs slightly (fibroglandular, $Z_{eff} = 7.4$, fat, $Z_{eff} = 5.9$). Breast cancer is found to have Z_{eff} that is quite close to fibroglandular tissue. The key to differentiating the normal fibroglandular tissue from the cancerous tissue is the slight difference in their linear attenuation coefficient at low photon energies (~17 to 25 keV). At lower photon energies, the photoelectric absorption is a dominant process, particularly with higher Z. Mammographic imaging makes use of this fact by using low kVp to obtain optimum tissue contrast (Figure 6.4).

a. Mo anode tube

Molybdenum anode tube has a $Z = 42$ and k-edge of 20 keV; therefore, the k-characteristic energy is ~17.3 keV. This means that a lot of X-ray photons are emitted at k-characteristic energy. Higher photoelectric absorption occurs in the cancerous tissue, making them appear whiter, thereby increasing sensitivity of cancer detection.

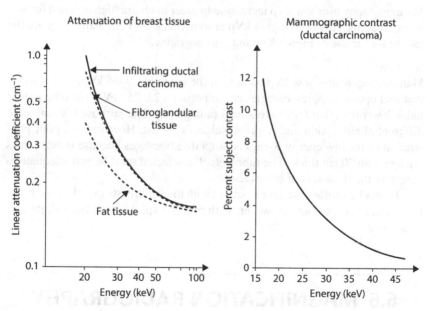

FIGURE 6.4 Linear attenuation of breast tissue and mammographic contrast for infiltrating ductal carcinoma. (Adapted from Yaffe, M. J., In Haus, A. G., Yaffe, M. J., [eds], *Syllabus: A Categorical Course in Physics: Technical Aspects of Breast Imaging*. Oak Brook, RSNA, 1994.)

b. W anode tube

Tungsten anode tube has a $Z = 74$ and k-edge of 69.5 keV; therefore, the k-characteristic X-ray energy is ~ 58.2 keV. The k-characteristic energy of tungsten is much higher and is not useful for the imaging of the soft tissues. However, the X-ray yield is much higher with a tungsten anode, which compensates for the extra attenuation and short exposure time. A Mo anode takes advantage of the k-characteristic radiation whereas the W anode tube uses the bremsstrahlung. The mean energy of a W anode tube is ~ 19 keV. When used in combination with a rhodium target, the advantage of the Rh absorption edge is still maintained.

6.5 HIGHER KVP TECHNIQUE

Problem:

Mammography uses low kVp technique in order to obtain high contrast for soft tissue imaging. However, higher kVp is sometimes used clinically. Explain the reason for the use of higher kVp in mammography.

Solution:

Mammography uses low kVp values, in the range of 20–35 kVp. It was found that that optimum X-ray energies range between 22–25 keV. This tube potential is lower than that for conventional radiography. The purpose is to increase differential absorption and improve subject contrast. However, the poor penetration of the low energy X-rays is not a disadvantages because most breasts are less than 10 cm thick. The tube potential selected must be just adequate to penetrate the thickness of breast tissue.

High kVp is often used for a patient with thicker breasts in order to achieve the required penetration. However, with higher kVp, there is also a slight loss of contrast.

6.6 MAGNIFICATION RADIOGRAPHY

Problem:

In mammography, magnification radiography is sometimes performed. Describe the principle of this technique. Discuss the advantage and disadvantage of magnification mammography.

Solution:

Magnification is used in mammography to examine suspicious areas of the breast. It improves the visualisation of the mass margins and fine calcifications. Magnification is achieved by moving the breast away from the detector/film, with a distance of 15–30 cm. The focus to detector distance (FDD) is kept constant, usually at 65 cm (Figure 6.5). The ratio of the focus to image detector distance (FDD) to the focus to object distance (FOD) gives the magnification (FDD/FOD). With a typical FDD of 65 cm, and FOD of 35 cm, the magnification is normally 1.85.

Advantages:

- Increased effective resolution of the image receptor by the magnification factor
- Reduced effective image noise
- Reduced scattered radiation reaching detector/film, eliminate the use of grid

Disadvantages:

- Limited field of view.
- Geometric unsharpness/blurring due to the finite focal spot size. This blurring is greater with larger focal spot size and higher magnification, resulting in a loss of resolution in the projected image. A 0.1 mm (small) focal spot is used to maintain resolution.

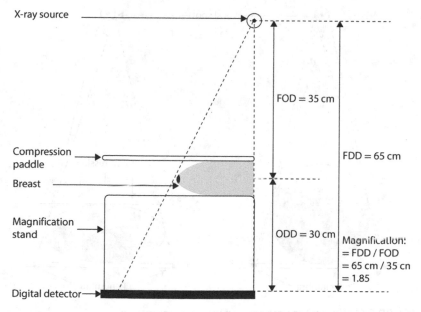

FIGURE 6.5 Schematic diagram of a mammography system.

6.7 COMPRESSION AND IMAGE QUALITY

Problem:
In mammographic examinations, the breast is compressed between two plates.

   a. Is the geometric unsharpness increased or decreased? Explain with an illustration.
   b. Explain how compression affects image contrast.
   c. Is the required radiation for a given image signal-to-noise ratio (SNR) higher or lower with compression? Explain briefly.

Solution:

   a. Geometric unsharpness decreases with the use of compression (Figure 6.6). Compression brings the subject closer to the detector, thus penumbra is decreased.
   b. Image contrast is improved due to less volume of tissue for scattering; the thickness traversed by the primary X-ray beam is also reduced. Less attenuation and less tissue volume leads to reduced beam hardening.
   c. For a given SNR, compression lowers radiation dose to the breast. Contrast is increased as noise, due to scattered radiation, is decreased.

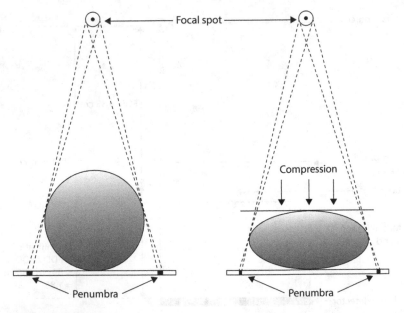

FIGURE 6.6 Geometric unsharpness due to compression.

6.8 MAMMOGRAPHY QUALITY CONTROL

Problem:
There is increasing emphasis today on quality control in mammography.

'Why do we single out mammography? Why don't we make sure chest radiography is being done correctly – there is more lung cancer than breast cancer and more chest radiographs are performed?'

Comment on the complaint above.

Solution:
Breasts intrinsically have low subject contrast that requires precise technique and positioning. The breast is also a very radiosensitive organ, thus strict quality control is needed to ensure optimal image quality and minimal dose is delivered to the breast. To perform high quality mammography requires dedicated trained personnel and the implementation of a quality control programme. On the other hand, chest radiography is relatively less stringent than mammography.

6.9 ANODE ANGLE

Problem:
Explain why an effective anode angle in the mammography X-ray tube should be at least 20°.

Solution:
Typical source to image distance for mammography equipment is between 60 cm to 70 cm. A focus to image distance (FID) of 65 cm requires the effective anode angle to be at least 20° to avoid field cutoff for the 24 cm × 30 cm field area. A shorter FID requires an even larger anode angle.

Fluoroscopy 7

7.1 IMAGE INTENSIFIER

Problem:

a. Figure 7.1 shows a schematic diagram of a typical image intensifier (II). Explain the function of the II with the aid of the labelled parts.

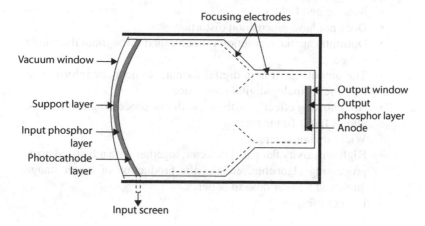

FIGURE 7.1 Schematic diagram of an image intensifier.

b. What advantages does a flat panel detector offer over an image intensifier tube in the fluoroscopy system?

Solution:

a. The input screen comprises of a vacuum window, a support layer, a layer of input phosphor and photocathode. The vacuum window keeps air out of the image intensifier (II) and usually has a curved shape to withstand outside air pressure. The support layer needs to be strong enough to support the subsequent input phosphor and photocathode layer, while being sufficiently thin to reduce X-ray attenuation. The housing commonly is made of 0.5 mm aluminium. The input phosphor layer absorbs X-ray photons and converts their energies to visible light. Caesium iodide (CsI) is a common material used. The photocathode layer is a thin layer of antimony and alkali material which emits electrons when struck by visible light.

The electrons produced by the photocathode material will undergo acceleration and focusing by the focusing electrodes, increasing their velocity and kinetic energy.

The output phosphor is located at the anode of the tube. It is made of zinc cadmium sulphide doped with silver (ZnCdS:Ag). The electrons impinging on the output phosphor produce green light emissions. The image produced at the output is an inverted and a minified version of the image at the input phosphor.

b. Advantages of flat panel detector over II:
   • Smaller and lighter.
   • Does not have pincushion distortion.
   • Optimum spatial resolution is maintained throughout the whole image.
   • The output signal is in digital format. No delay or information loss due to analog-digital conversion.
   • No 'ghosting effect', combined with fast processing algorithms, allows faster frame rates.
   • Wider dynamic range.
   • High sensitivity flat panel detector, together with advanced post-processing algorithm, enables the production of better image quality at a lower dose to patients.
   • Longer lifespan.

Problem:
How do the sizes of the input and output screens affect the performance of an image intensifier?

Solution:
The output screen is smaller than the input screen. This results in the minification of the output image but the brightness of the output image is increased by a factor equal to the ratio of the input and output screen areas. For example, given the area of the input screen is 25 cm^2 and the output screen is 2.5 cm^2, the brightness of the output image is increased by a factor of 100.

7.2 FLUORESCENT SCREENS

Problem:

Discuss the use of fluorescent screens in radiography and fluoroscopy.

Solution:

In radiography – fluorescent screens are used in the intensifying screens. For general radiography, the conventional material used is $CaWO_4$ but nowadays, the commonly used materials are Gd_2O_2S, LaOBr and $YTaO_4$. Gd_2O_2S:Tb are also used for mammographic screens. The fluorescent screens are coupled to the film. They produce light photons, which are more efficient in darkening the film. This is because one X-ray photon is only able to sensitise one silver halide grain. However, the same X-ray photon can produce many 2–3 eV light photons; therefore, it is able to sensitise many more silver halide grains in the film.

In fluoroscopy – fluorescent screens are being used at the input screen and the output screen of the image intensifier. CsI:Na is the common material that is used for the input screen and it is coupled to a layer of photocathode, which will produce electrons. On the output screen, ZnCdS:Ag is often used to convert electrons to light photons, which are then coupled to a camera system for image display and readout.

Note: For further readings, please refer Ng et al., 2017.

7.3 CHARGE-COUPLED DEVICE

Problem:

Describe a charge coupled device (CCD) and how it functions to produce a digital image in a fluoroscopic imaging chain.

Solution:

Charge coupled device (CCD) is a type of semiconductor optical imaging scanner. CCD sensors are usually made of etched amorphous silicon (aSi) wafer, resulting in an array of pixel elements (~50 µm in size). They come in a variety of matrix sizes, ranging from 1024 × 1024 to 4096 × 4096. The physical size of a CCD usually ranges from 5 to 20 cm. Each pixel element acts as a capacitor that will store electrical charges until it reaches a predetermined bias voltage.

CCD functions as an indirect X-ray detector. It is usually coupled to a fluorescent screen or a photoelectric cathode surface. When exposed to radiation, X-ray photons will produce light photons in the fluorescent screen. The light photons subsequently will be converted to electrons by the amorphous

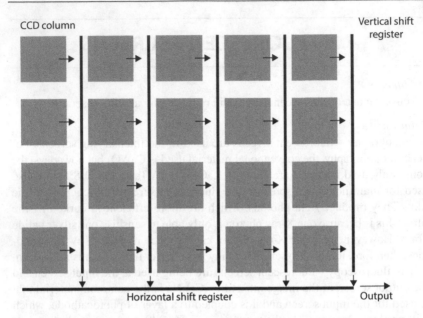

FIGURE 7.2   Shift register method.

Si elements in the CCD. The electrical charge produced is proportional to the total number of X-ray photons.

To read out, the CCD employs a shift register method (Figure 7.2). The electrical signal is passed to the analog to digital converter and digitized. It is then placed in a pixel array generated in the computer for viewing.

7.4 ARTEFACT

Problem:
Figure 7.3 shows a pin cushion artefact that can occur with image intensifiers (II) for use in fluoroscopy. Explain briefly what causes this artefact, and state one method that can be used to avoid/reduce this artefact.

Solution:
The pincushion distortion is caused by the physical shape differences of the input and output phosphor of the image intensifier. The input phosphor is curved to allow for electron focusing, while the output phosphor is flat, resulting in the warped output image. Pincushion distortion is reduced when magnification modes are used.

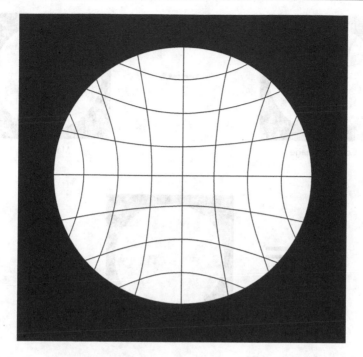

FIGURE 7.3 Pin cushion artefact.

7.5 DIGITAL SUBTRACTION ANGIOGRAPHY

Problem:

Explain how digital subtraction angiography (DSA) works and give examples where this technique is used clinically.

Solution:

Digital subtraction angiography (DSA) is a subtraction technique involving two sequential images acquired at the same position. It is accomplished by taking into account the changes in the linear attenuation coefficients of the imaged object/body parts due to the presence of contrast medium. The principle of DSA imaging is illustrated in Figure 7.4.

An image is first acquired prior to the injection of contrast medium. This image is called the 'mask' image. Subsequently, contrast is injected into the

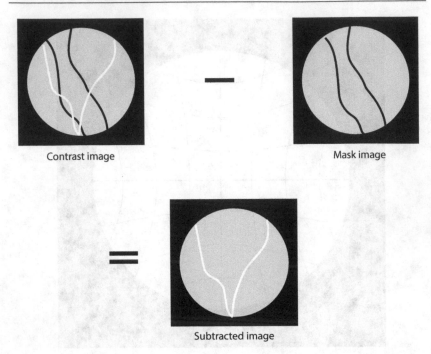

Contrast image

Mask image

Subtracted image

FIGURE 7.4   Principle of DSA.

blood vessels of the patients. Another image is acquired during this time, and it is called the 'contrast' image. The 'mask' image is then superimposed on the 'contrast' image and subtracted from the image.

DSA is often used to image blood vessels. It is a useful technique to subtract anatomical image noise such as soft tissue and anatomical structures. The position of the blood vessels is highlighted by the contrast medium in the blood vessels.

7.6 NOISE

Problem:
Digital subtraction angiography (DSA) is one of the most often used techniques during interventional radiology procedures. Comment on how digital image subtraction affects quantum noise and structure noise in a digital image. Suggest a method to reduce quantum noise in DSA images.

Solution:

Digital image subtraction increases image noise. This is also known as 'quantum noise', which are random fluctuations of image signals. Quantum noise increases as the signal photons in the images decrease. It is represented by the Poisson error ($\sqrt{n}$), where n is the number of signal photons on the image. On the other hand, 'structure noise' are anatomical structures that hinder proper visualization of blood vessels of interest. DSA actually works by suppressing/subtracting anatomical details (or structure noise).

However, this procedure reduces the image signal and therefore increases quantum noise of the image. One method to reduce quantum noise in fluoroscopic images via post-processing techniques is to apply a temporal frame summing technique, i.e. sequential digital image frames are summed retrospectively. For example, five frames of mask images are summed together and subtracted with five frames of post contrast injection. This technique will increase signal-to-noise ratio of the DSA image without increasing radiation dose to patient.

7.7 BIPLANE ANGIOGRAPHY

Problem:

Compare and contrast the advantages and disadvantages of using a biplane angiography unit.

Solution:

Advantages:

- Simultaneous imaging of vascular structures in two different planes.
- Two projections of the anatomy can be obtained with a single injection of a contrast medium. This may potentially reduce the use of contrast medium.
- The overall duration of the procedure can be shortened.
- A biplane system provides full body coverage for neurovascular and vascular interventional procedures.

Disadvantages:

- Scattered radiation from the second tube tends to degrade contrast of the angiographic image.
- Biplane angiography units are more difficult to operate because they require appropriate selection of technical exposure factors for the simultaneous biplane projections.
- Requires appropriate selection of technical exposure factors.
- Higher equipment and installation and maintenance cost.

7.8 POST-PROCESSING TECHNIQUES

Problem:

The modern fluoroscopy and angiography unit relies heavily on post-processing techniques in order to reduce radiation dose or contrast to patient. Explain how the following post-processing techniques works:

    a. Temporal frame averaging
    b. Edge enhancement
    c. Last image hold
    d. Roadmapping

Solution:

    a. **Temporal frame averaging** – is a technique to decrease the noise level in the displayed image. Fluoroscopic images are relatively noisy due to the low tube exposure. However, when a few sequential fluoroscopic images were added together and averaged, the resulting image would be less noisy. Averaging more frames together decreases image noise further. However, too much temporal filtering also causes noticeable image lag.

    b. **Edge enhancement** – is a technique to increase the detectability of small objects and object edges. Edge-enhanced images are produced by first creating a blurred version of the original image. This is usually done by averaging a pixel value with surrounding pixel values and placing the new value in the location of the original pixel (e.g. averaging the nine pixels in a 3×3 kernel and placing the averaged value at the centre). The blurred image is then subtracted from the original image. This produces an 'edge image'. The edge image is then added onto the original image to produce the edge-enhanced image.

$$\text{Edge-enhanced image} = \left(\begin{array}{c} \text{original} \\ \text{image} \end{array} - \begin{array}{c} \text{blurred} \\ \text{version} \end{array} \right) + \text{original image}$$

    c. **Last image hold** – uses the digital information from the last stored frame and continuously displays it on the monitor, showing the patient's anatomy even after the X-rays beam has been turned off. This technique decreases radiation exposure to the patient and radiology staff.

    d. **Road mapping** – is a technique that facilitates the placement of devices, such as stents and balloons, assisting the navigation of guide wires and catheters through severe stenosis and tortuous vessels. This technique also reduces the total contrast medium and radiation dose to patients. Two different approaches to road mapping can be found on fluoroscopic systems:

- Reference and live image displayed on side by side monitors
- Reference image overlay onto the live image displayed on the monitor

7.9 FLUOROSCOPIC IMAGE QUALITY

Problem:
Explain how the following will alter the fluoroscopic image quality:

    a. Field of view (FOV)
    b. Automatic brightness control (ABC)
    c. Frame rate

Solution:

    a. Smaller field-of-view produces better details. This is usually achieved by collimation and increased exposure. Hence, it will increase the patient dose.
    b. The use of ABC keeps the image brightness of a II system regardless of the patient size.
    c. Using higher frame rates will reduce the flickering and motion artefact.

7.10 PATIENT DOSE REDUCTION

Problem:
Describe several methods to minimise radiation dose to patient during a fluoroscopy procedure.

Solution:
Methods to minimise radiation dose to patients during a fluoroscopy procedure are:

- Minimise beam-on time
- Maximise source to patient distance
- Optimize kVp
- Collimate to region of interest
- Use last image hold and pulsed fluoroscopy
- Review real time image recordings
- Minimise use of magnification modes

7.11 KERMA AREA PRODUCT (KAP)

Problem:
Explain the dose quantity kerma area product (KAP) used in the monitoring of the X-ray output of fluoroscopy. Discuss the relationship of KAP with the varying of SID.

Solution:
Kerma area product (KAP) is air-kerma at a point along the central axis of the X-ray beam, multiplied by the exposure field size at that point (mGy.cm²). KAP is measured by a transmission plane parallel ionisation chamber that is placed at the output of the X-ray tube, between the collimator and patient's body.

KAP quantifies the radiation output of the X-ray tube for a fluoroscopy procedure. However, it is often used as a dose metric to monitor the radiation exposure to patient, and to approximate stochastic risk to patients due to radiation received during fluoroscopy procedures. KAP provides no information regarding the radiation dose distribution over the patient's body. The same KAP is observed with small field size and high radiation dose with large exposure field size and low radiation dose (Figure 7.5).

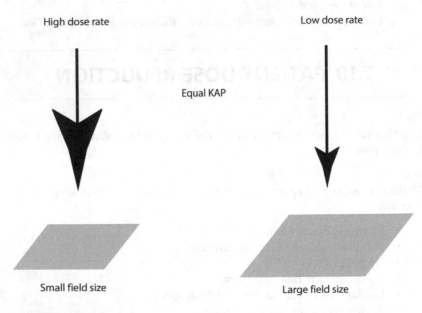

High dose rate   Low dose rate

Equal KAP

Small field size   Large field size

FIGURE 7.5   KAP.

7.12 INTERVENTIONAL REFERENCE POINT

Problem:

Explain the interventional reference point (IRP) that was defined by the International Electrotechnical Commission (IEC). Discuss the purpose and the usefulness of this dose quantity for patient dose monitoring during fluoroscopy procedures.

Solution:

The interventional reference point (IRP) was introduced by the International Electrotechnical Commission (IEC) in 2000. It is defined to be located at 15 cm from the isocentre of the C-arm toward the X-ray tube (IEC, 2000) (Figure 7.6). This point is fixed relative to the X-ray tube and is independent of the patient variation.

The IRP is fixed relative to the X-ray tube as it is always at a constant distance from the isocentre, which may represent a position on the patient's skin or a point inside or outside of the patient's body (Figure 7.7). The IRP rarely

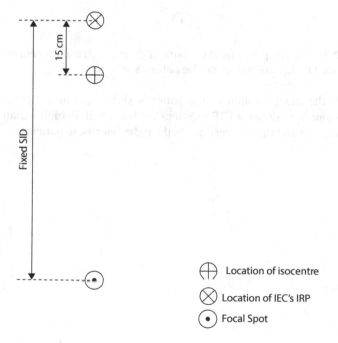

FIGURE 7.6   The IRP and FDA defined reference dose points.

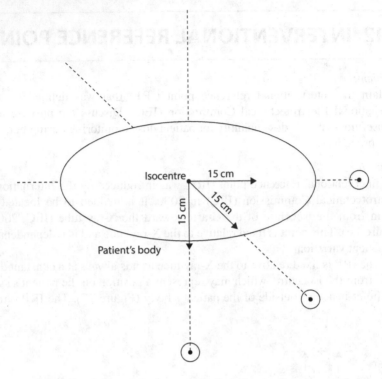

FIGURE 7.7 The IRP positions (IEC method) during a clinical procedure, which can be placed inside, outside, or on the patient's skin.

represents the exact location of the patient's skin over clinical procedures. Using accumulated doses at IRP locations tends to result in higher than actual peak skin doses and thus overestimates the radiation risk to patients.

Computed Tomography 8

8.1 SPIRAL CT

Problem:

What is spiral (or helical) CT scanning? Explain briefly the technology that enables spiral CT scanning.

The main advantage of spiral scanning is the reduction of image acquisition time. State four possible advantages related to this.

Solution:

With spiral CT scanning, the patient is moved continuously through the gantry during the CT study. This is in contrast to the shoot and move technique in the earlier generation CT scanner. The spiral CT technology is made possible with the invention of the slip ring technology. A slip ring technology has a circular contact with sliding brushes that allows the gantry to rotate continually around the CT gantry. It is untethered from wires.

Four advantages following the reduction of image acquisition time include,

- Reduction of patient motion artefact
- Increase in patient throughput
- Reduction in the volume of contrast medium needed
- Absence of inter-slice gaps improves the quality of 3D reconstruction

---

8.2 NOISE

Problem:

A series of 10 CT slices with 2 mm thickness were acquired at 120 kVp, 1 second, and 100 mA. The images were considered to be too noisy. Suggest an adjustment to the technique in order to double the signal-to-noise ratio (SNR).

Solution:
There are two ways to increase the SNR by a factor of 2.

Solution #1 is to increase the tube current-time product (mAs), i.e. using the parameter 120 kVp, 100 mA, 4 seconds, 2 mm thick.

Solution #2 is to increase the scan slice thickness, i.e. 120 kVp, 100 mA, 1 second, 4 mm thick.

The signal in the CT image is proportional to the number of detected X-ray photons obeying Poisson statistics. Therefore, the uncertainty of the signal or 'noise' is the square root of the detected X-ray photons.

Therefore,

$$\text{SNR} = \frac{\text{Signal}}{\sqrt{\text{signal}}}$$

To increase the SNR by a factor of 2,

$$\frac{\sqrt{S_2}}{\sqrt{S_1}} = 2$$

$$\frac{S_2}{S_1} = 4$$

The number of detected X-ray photons must be 4x that of the initial signal. Since the number of X-ray photons is directly proportional to the tube current (mAs), and initially 100 mAs, 100 mA × 4 sec = 400 mAs would result in a doubling of the SNR.

Note that for CT imaging, the signal is actually the difference in X-ray attenuation between adjacent tissues. Thus, signal represents the difference in the number of X-ray photons detected that give rise to the signal (contrast) in X-ray imaging. However, the above equation holds for both cases.

For the second solution, based on the same concept that to achieve a doubling of the SNR, the number of photons detected must be quadrupled. The number of detected X-ray photons is linearly proportional to the slice thickness. A combination of doubling the mAs and the slice thickness will result in a quadrupling of the number of detected X-ray photons.

Problem:
A CT scan that employs a 10 mm slice width with the scan passing symmetrically between the endplates of the vertebral body is performed on a patient's lumbar vertebrae (for a bone density study). A region of interest (ROI) is drawn over a trabecular bone of the vertebral body. The mean HU of the ROI is 100 HU with an associated standard deviation (SD) of 15 HU. The number of pixels enclosed by the ROI is N = 900. The measured mean pixel value is

assumed to represent the trabecular bone density. What do the mean and SD of the ROI mean?

Solution:

Approximately 68% of the enclosed pixels have values ranging between 85 and 115 HU.

In most cases with ROIs placed over homogenous tissue (assuming trabecular bone really is homogenous tissue), the distribution of pixel values follows a normal distribution for a homogeneous tissue according to the Central Limit Theorem. The width of the curve or the degree of scatter of the pixel values is described by the SD. In such a distribution, 68% of the pixel values fall within ±1 SD of the mean, i.e. there is a 68% chance that a pixel will have a value within that range. 95% of the pixel values fall within ±2 SD of the mean. If two ROIs with different mean pixel values are compared, there will be a 95% chance that two tissues actually have different attenuation properties if their mean values differ by 1.4×2 SD.

Thus, the mean pixel values for the two patients would have to differ by at least $2.8 \times 15 = 42$ HU to be interpreted as different with 95% confidence. In many successive CT scans and ROI measurements on the same patient, the distribution of the pixel value means will also be characterised by a SD, but in this case, it is generally referred to as a 'standard error' (SE), where

$$SE = \frac{SD}{\sqrt{N}}$$

In the present case,

$$SE = \frac{15}{\sqrt{900}} = 0.5\,HU$$

Hence, the SE of the mean pixel values would be 0.5 and not 15 HU. The SE represents the uncertainty in the mean values. 68% percent of the mean values would fall between ±1 SE.

Note:

The 1.4 is derives from a statistical analysis of the two Poisson-distributed pixel populations.

8.3 CT NUMBER

Problem:

a. Define CT number.
b. How does the CT number vary with tube voltage (kV) on a CT scanner?

If linear attenuation coefficients for fat and water are 0.17 cm⁻¹ water 0.19 cm⁻¹, respectively, calculate the CT number for fat.

Solution:

a. The CT number is defined by the linear attenuation coefficients, μ of tissues and water, expressed as Hounsfield units (HU).

$$\text{CT number}_{\text{tissue}} = 1000 \frac{\mu_{\text{tissue}} - \mu_{\text{water}}}{\mu_{\text{water}}}$$

b. As the tube voltage (kV) increases, the X-ray beam energies become harder and more penetrative; the linear attenuation coefficients of tissues decrease (Figure 8.1). The key is the difference between the linear attenuation coefficients of tissues with water (Figure 8.2). Bony tissues and contrast material, such as iodine and barium, have relatively higher effective atomic numbers compared to soft tissues and lungs. At the energy range commonly used in diagnostic CT, the probability of photoelectric absorption has a large effect on the linear attenuation coefficients.

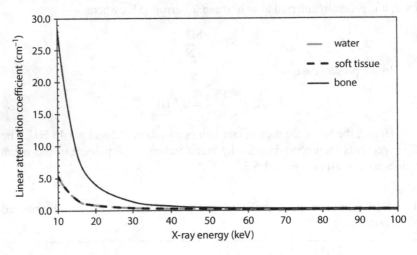

FIGURE 8.1 Linear attenuation coefficient of water, soft tissue and bone as a function of X-ray energy.

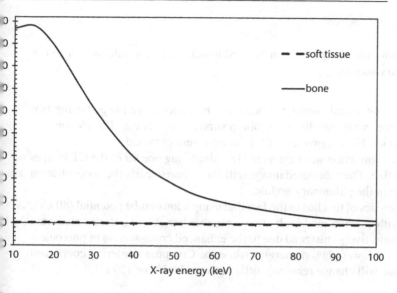

E 8.2 CT number of soft tissue and bony tissue as a function of X-ray

1e net effect is shown in Figure 8.2; as the X-ray energy increases, the CT
r of tissues decreases. However, the rate of decrease of the bone is much
and both soft tissue and bone will appear darker on the CT image.
1e CT number for fat is

$$\text{CT number}_{\text{fat}} = 1000\frac{0.17 - 0.19}{0.19} = -105.2\,\text{HU}$$

m:
ins were performed on a 16 cm PMMA cylinder using 100 kV and
. Which tube voltage is expected to produce a higher CT number of the
r?

n:
'p will produce a higher CT number.
number is proportional to μ of the PMMA cylinder (from CT num
ation), and μ is inversely proportional to photon energy, i.e. spectrum
e of less attenuation from higher energy based on the exponential
tion formula.

8.4 DUAL ENERGY CT

Problem:
Explain how dual-energy CT can be used to determine the calcium content in a pulmonary nodule.

Solution:
When utilising a dual energy CT scanner, the patients are imaged using two X-ray sources with two different tube potentials, e.g. one at 120 kV and the other at 80 kV. Hence, two sets of CT images were obtained.

Image subtraction was performed by subtracting one set of the CT images from the other. The subtracted image will show more clearly the concentration of calcium in the pulmonary nodule.

The physics of this lies in the fact that using a lower tube potential (80 kV), nodules with calcium (and high atomic number bones and contrast medium), would be selectively enhanced due to the enhanced cross-section of photoelectric effects at lower photon energies while the Compton scattering coefficient of the tissue will change relatively little between 80 kV or 120 kV.

8.5 PARTIAL VOLUME EFFECT

Problem:
Explain partial volume effect in CT. With an illustration, explain the 'partial volume' effect. How could this effect be reduced?

Solution:
Partial volume is formed when high attenuating objects (e.g. bone) project partly into a slice that contains a low attenuating object (e.g. soft tissue) causing a mixed attenuation value. The attenuation coefficient that will be calculated for that voxel will be a weighted average of that for soft tissue and bone based on the relative volumes occupied by both in the voxel.

Figure 8.3 illustrates partial volume effect. On the right side, the X-ray beam only scanned part of the object, resulting in partial volume effect whereby the averaging of the CT number of the soft tissue around the dense object produces a lower CT number. On the left side, the X-ray beam was transmitted through the dense object, resulting in a higher CT number.

In other words, partial volume effect occurs where the object does not fill the scan plane. CT number is therefore underestimated.

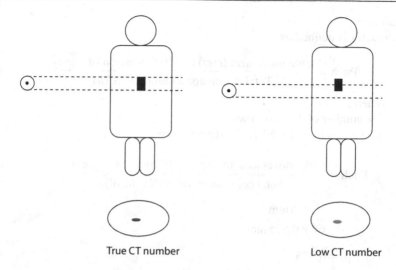

True CT number   Low CT number

FIGURE 8.3 Partial volume effect.

Ways to reduce the partial volume effect:

- Reducing FOV
- Reducing slice thickness
- Overlapping slices

8.6 PITCH

Problem:
Define pitch factor for a multi-slice, helical CT.

    a. Calculate the pitch of a 64 slice CT, given that the table moves 15 mm per gantry rotation, and the total detector collimator width is 64×0.625 mm.

    b. Differentiate between the spirals created using pitch = 1, 2 and 0.5 in terms of their coverage of the region of interest in the patient.

    c. How is the image quality and patient dose affected if the pitch is increased?

    d. The advantage of using low pitch is improved spatial resolution along the Z-axis of the patient. State two of its disadvantages.

Solution:
Pitch factor is defined as,

$$\text{Pitch} = \frac{\text{table movement (mm) per } 360° \text{ rotation of gantry}}{\text{Total collimator width (N} \times \text{hCol)}}$$

Where,
N = number of detector rows
hCol = height of an individual detector row

$$\text{Pitch} = \frac{\text{table movement (mm) per } 360° \text{ rotation of gantry}}{\text{Total collimator width (N} \times \text{hCol)}}$$

a. $= \dfrac{15\,\text{mm}}{64 \times 0.625\,\text{mm}}$

$= 0.375$

b. A pitch of one yields a contiguous spiral. A pitch of two yields an extended spiral. A pitch of 0.5 yields an overlapping spiral.
c. When the pitch increases, the image quality will be degraded because of lower SNR due to partial scanning effects, i.e. each image slice does not consist of a full 360° scan. The radiation dose will be reduced as well because the scan time is reduced for the same coverage area.
d. The two disadvantages are higher patient doses and longer imaging times.

8.7 TABLE SPEED

Problem:
Calculate the table speed for a 32 slice collimation with 0.6 mm slice width and 0.4 s rotation time at a pitch of 1.5.

Solution:

$$\text{Table speed} = \frac{\text{pitch} \times \text{slice collimation} \times \text{slice width}}{\text{rotation time}}$$

$$= \frac{1.5 \times 32 \times 0.6\,\text{mm}}{0.4\,\text{s}}$$

$$= 72\,\text{mm s}^{-1}$$

8.8 SPATIAL RESOLUTION

Problem:

   a. List five factors that affect the spatial resolution of a CT scanner.

   b. Calculate the pixel size for a 40 cm FOV in a standard 512 × 512 image matrix.

Solution:

   a. The spatial resolution of a CT scanner depends on a number of factors:

- Number and size of individual detector elements
- Slice thickness
- Focal spot size
- Image matrix
- Field of view (FOV)
- Reconstruction algorithm

   b. Calculation of pixel size

$$\text{pixel size} = \frac{\text{FOV}}{\text{image matrix}}$$

$$= \frac{400\,\text{mm}}{512}$$

$$= 0.78\,\text{mm}$$

Magnetic Resonance Imaging

9

9.1 NUCLEAR SPIN

Problem:

What is the nuclear spin of a hydrogen-1 (^{1}H) nucleus? Explain nuclear spin.

Solution:

$$\text{Spin}(^1\text{H}) = \frac{1}{2}$$

The magnetic property of a nucleus derives from a quantum mechanical property called 'spin', which is a property inherent in charged particles. Quantum spin is analogous to, but not the same thing as, the classical concept of spin. Since they are quantised, spins for individual protons and neutrons must have discrete values. This value happens to be the same as for the electron spin; i.e. spin = ½. The total spin of a particular nucleus (I) is determined by the sum of the spins of its protons and neutrons. Since a ^{1}H nucleus has only a single proton, its spin must be equal to ½.

Problem:

Does a carbon-12 (^{12}C) nucleus have spin? Explain your answer.

Solution:

No. The shell model of nuclei posits that protons or neutrons will form in pairs, each constituent having opposite total angular momentum. The spins of paired nucleons cancel (+½ + −½ = 0) so that the magnetic moment of a nucleus with even numbers of both protons and neutrons is zero. Magnetic nuclei, therefore, are those with an odd number of protons and/or odd number of neutrons. These magnetic nuclei will have at least one unpaired proton or one unpaired neutron.

From this discussion, we can conclude that all nuclei with an even atomic number (Z) and even neutron number (N) have nuclear spin $I = 0$. Thus, the ^{12}C nucleus cannot have spin. However, the nuclear magnetic moment varies from isotope to isotope of an element, thus both the ^{11}C and ^{13}C isotopes do have spin.

Problem:
Can an uncharged particle, like a neutron, have nonzero spin?

Solution:
Yes. The neutron is composed of other, electrically charged particles called quarks. Quarks have fractional electric charge values. 'Up' quarks have a charge of +2/3, while 'Down' quarks have −1/3 charge. Quarks are also classified as spin-1/2 particles. Protons consist of two 'Up' quarks and one 'Down' quark. Neutrons are composed of two 'Down' quarks and one 'Up' quark. The 'Up' and 'Down' quarks in each nucleon will pair so their spins will cancel. Thus, the total spin of a neutron is ½ (+ ½ − ½ + ½ = ½).

9.2 PRECESSION

Problem:
What is the type of motion called 'precession'?

Solution:
It is the motion manifested as the axis of a spinning object describes a cone in space when an external torque is applied to it.

Classically, the magnetic dipole moment is a measure of the magnetic strength of a magnet or current-carrying coil, described in terms of the torque that the magnet experiences when placed in an external magnetic field (see Figure 9.1). As the electrons travel around the coil, the north pole of the magnet is attracted to the south pole of the external magnetic field, causing a net rotation of the current loop. The magnetic moment has a close connection with angular momentum through a phenomenon called the 'gyromagnetic effect'. The 'gyromagnetic effect' manifests when a magnetic moment is subject to a torque in a magnetic field that tends to align it with the applied magnetic field. The moment 'precesses', i.e. it rotates about the axis of the applied field. This is the same motion that a gyroscope makes as it is spinning and rotating within a gravitational field.

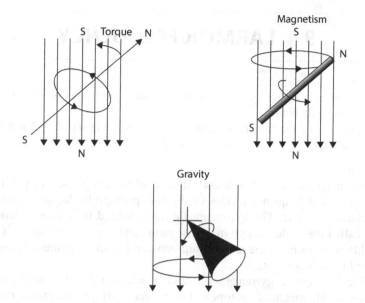

FIGURE 9.1 Precession.

9.3 GYROMAGNETIC RATIO

Problem:

What is the gyromagnetic ratio (γ) and what is the γ of the hydrogen-1(^{1}H) nucleus?

Solution:

The γ of the hydrogen-1(^{1}H) is 42.576 MHz T^{-1}.

It is this magnetic moment of the nucleus that allows the observation of nuclear magnetic resonance (NMR) absorption spectra caused by transitions between nuclear spin levels. For any magnetic nucleus, the ratio of the magnetic moment to the angular momentum is known as gyromagnetic ratio (γ). γ is a scalar quantity since the magnetic moment vector is parallel to angular momentum vector. The value of γ is unique for each magnetic nuclear isotope. (All non-magnetic nuclei have gamma = 0). The hydrogen-1 gyromagnetic ratio, $\gamma = 42.576$ MHz T^{-1}, is the greatest of all isotopes.

9.4 LARMOR FREQUENCY

Problem:
Describe what is meant by the Larmor frequency and how it is related to the applied magnetic field.

Calculate the Larmor frequencies of protons in MHz at 0.35, 1.5 and 3.0 T. Given that the gyromagnetic ratio of protons is 42.6 MHz T^{-1}.

Solution:
The gyromagnetic ratio, γ, is defined in units of radians per second per Tesla. We express it in frequency (cycles s^{-1}) by multiplying it by 2π, since there are 2π radians per cycle. The gyromagnetic ratio, defined in frequency units, is often called the 'reduced gyromagnetic ratio', and is given the sign γ. (Note that this is the same as the relationship between Planck's constant, h, and the reduced Plank's constant, $\hbar$.)

The hydrogen-1 gyromagnetic ratio, $\gamma = 267,513$ $rad^{-1}s^{-1}mT^{-1}$, is the greatest of all magnetic isotopes. The reduced 1H gyromagnetic ratio is $\not\gamma = 42,576$ Hz mT^{-1}.

The gyromagnetic ratio is also important because it is the constant that describes the 'gyromagnetic effect'. The 'gyromagnetic effect' is summarised by the Larmor equation, which relates the frequency of precession to the strength of the applied magnetic field:

$$\nu_o = \not\gamma \, B_o,$$

where,
ν_o is the resonant or 'Larmor' frequency (in Hz)
B_o is the applied magnetic field (in T).

Larmor frequencies are 14.9 MHz, 63.9 MHz and 127.8 MHz for 0.35 T, 1.5 T and 3.0 T, respectively.

9.5 MRI IMAGES

Problem:

a. Label the following MR images (see Figure 9.2).
b. Based on the saturation recovery plot (Figure 9.3), comment on the MR image that you would expect to see.

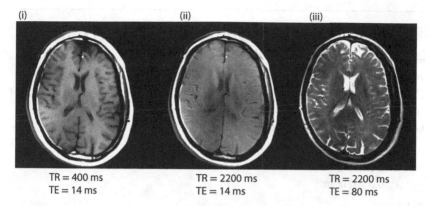

(i) (ii) (iii)

TR = 400 ms
TE = 14 ms

TR = 2200 ms
TE = 14 ms

TR = 2200 ms
TE = 80 ms

FIGURE 9.2 MR images for different sequences.

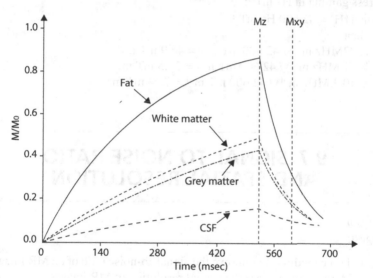

FIGURE 9.3 Saturation recovery plot.

Solution:

a. MR image types:
   i. T_1-weighted (short TR, short TE)
   ii. Proton density (long TR, short TE)
   iii. T_2-weighted (long TR, long TE)

b. This image will have very low contrast; in particular, there will be almost no T1-based contrast between white and grey matter. There is also proton density difference between white matter (~75%) and grey matter (~85%), which influences T1-weighted image contrast.

9.6 LONGITUDINAL RELAXATION

Problem:
Convert the following gradient strengths, which are expressed in terms of frequency, into mT m^{-1}.
Given the gyromagnetic ratio = 42.6 MHz T^{-1}.

    a. 2 MHz m^{-1}
    b. 1 MHz m^{-1}
    c. 0.3 MHz m^{-1}

Solution:
Express gamma in Hz mT^{-1}:
gamma(1H) = 42 600 Hz mT^{-1}
    then
    a. (2MHz m^{-1}) / 42 600 mT m^{-1} = 46.9 mT m^{-1}
    b. (1 MHz m^{-1}) / 42 600 mT m^{-1} = 23.5 mT m^{-1}
    c. (0.3 MHz m^{-1}) / 42 600 mT m^{-1} = 7.04 mT m^{-1}

9.7 SIGNAL TO NOISE RATIO AND SPATIAL RESOLUTION

Problem:
Suggest:

    a. Three methods to improve the signal-to-noise ratio of an MR image
    b. Methods to increase the spatial resolution of MR images

Solution:

    a. The signal-to-noise ratio of an MR image can be improved by:
- Increasing voxel size
- Increasing slice thickness
- Using scanner with higher magnetic field, i.e. 7 T scanner has higher signal compared to 3 T scanner

b. Methods to increase the spatial resolution of MR images
- Use thinner slices.
- Use smaller field of view.
- Increase matrix size.

9.8 MAGNETIC FIELD GRADIENT

Problem:

Figure 9.4 (a) shows the local RF signal received by the RF coils due to the three different volumes of water in a homogeneous magnetic field of 1.5 T. During a slice selection process, the field gradient of 10 mT m^{-1} is switched on, resulting in a field gradient as shown in Figure 9.4b. Sketch the expected local RF signal received by the RF coils.

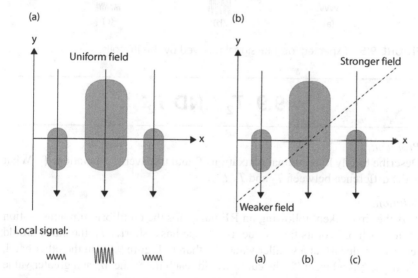

FIGURE 9.4 (a) Local RF signal received by the RF coils due to the three different volumes of water in a homogeneous magnetic field of 1.5 T and (b) when a magnetic field gradient was applied onto the same objects.

Solution:

Figure 9.5 shows the expected local RF signal received by the RF coils.

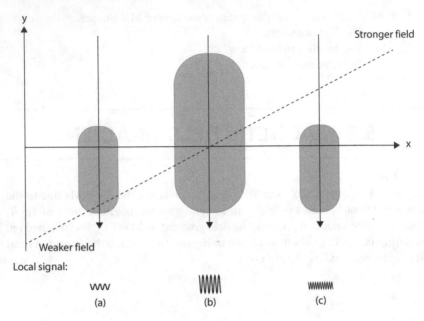

FIGURE 9.5   Expected local RF signal received by the RF coils.

9.9 T_2 AND T_2^*

Problem:
Describe briefly longitudinal relaxation, T_1 and transverse relaxation, T_2. What is the difference between T_2 and T_2^*?

Solution:
T_1 is the time taken, following an RF pulse, for the equilibrium magnetisation M_z to reach 63% of its final value. If a tissue has a shorter T_1, the curve would reach the value M_o at a smaller value of t than in Figure 9.6. On the other hand, if a tissue has a longer T_1, the curve would reach the value M_o at a greater value of t than shown in the figure.

Transverse relaxation is the time taken for the loss of signal to 37% due to dephasing of the traverse magnetisation (Figure 9.7). It is mainly due to spin-spin interactions (random interactions with surrounding matter) and small inhomogeneities of the external magnetic field causing the protons to dephase. T_2 effects are due to molecular spin-spin interactions, which are irreversible. The inhomogeneity effects are additive to the molecular effects, resulting in a faster decrease in the MRI signal due to dephasing of the transverse macroscopic inhomogeneities is known as T_2^*.

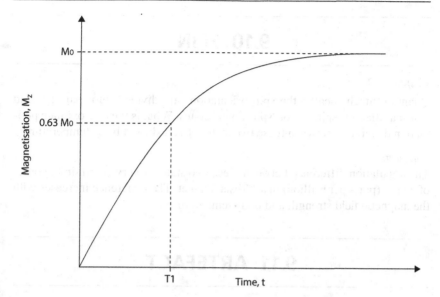

FIGURE 9.6 Longitudinal relaxation, T_1.

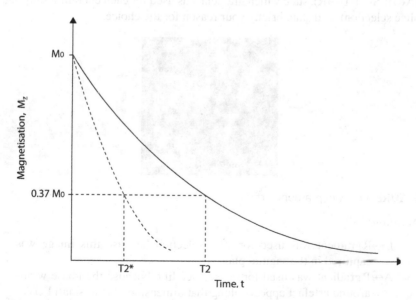

FIGURE 9.7 Transverse relaxation, T_2.

9.10 SPIN

Problem:
Quantum mechanically, the spin orientations are divided into 'parallel' and 'anti-parallel' or 'spin-up' or 'spin-down' states. What is the approximate population difference between these two states at 1 Tesla and body temperature?

Solution:
The population difference between these two states is very small, in the order of 6 ppm (parts per million) in a 1 Tesla magnet. The difference increases with the magnetic field strength and body temperature.

9.11 ARTEFACT

Problem:
Figure 9.8 is an example of the aliasing or wrap-around artefact. It is related to spatial encoding. Given that there are three separate spatial encoding gradients (A↔P, S↔I, L↔R), state which gradient was used for each encoding step (e.g. slice selection), and state briefly your reason for the choice.

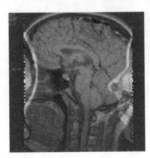

FIGURE 9.8 Wrap-around artefact.

Solution:

- L↔R gradient was used for slice selection because this image was captured in the sagittal plane.
- A↔P gradient was used for phase encoding because the phase wrap-around artefact appears along that dimension, due to small FOV.
- S↔I gradient was used for frequency encoding because no wrap-around artefact appears along that dimension, even though there is tissue outside the FOV (below the neck).

9.12 T_2^*

Problem:
Figure 9.9 depicts the theoretical T_2 transverse magnetisation relaxation (decay) after a 90° RF pulse. Redraw the graph and add in the T_2^* curve, using the diagram to explain why T_2^* exists, and how the spin echo pulse sequence could be used to make a 'true' T_2 measurement. Include in your diagram TE and any necessary RF pulses.

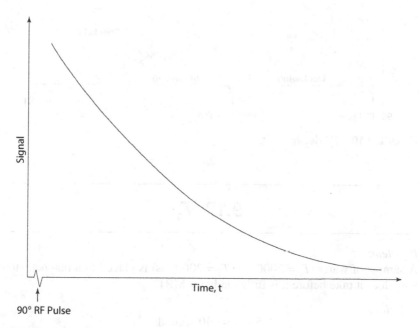

FIGURE 9.9 T_2 decay.

Solution:
Answer shown in Figure 9.10.

'True' T_2 should only reflect relaxation due to spin-spin interactions / tissue characteristics. T_2^* exists due to field inhomogeneities / susceptibility causing faster transverse decay. A 180° RF pulse is applied at time ½ TE to flip the dephased magnetisation to the other side of the transverse plane. Although flipped and still dephased, the transverse magnetisation still experiences the same inhomogeneities and thus gradually will rephase again. After another ½ TE, the rephasing will be complete and the signal (echo event) is measured.

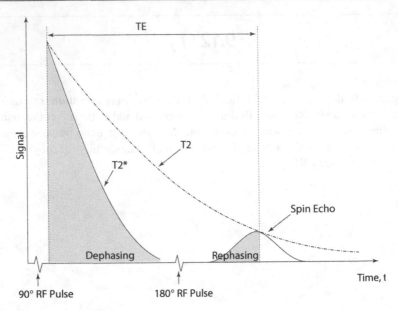

FIGURE 9.10 T_2^* decay.

9.13 T_1

Problem:
A sample of water ($T_1 = 2000$ ms, $T_2 = 2000$ ms) is placed in a magnet. How long does it take before it is fully ready for MRI?

Solution:
$$5 \times T_1 = \sim 10 \text{ seconds}$$

The time constant, T_1, is called the longitudinal relaxation time because it characterises how long it takes for the nuclear spins to become fully aligned with the B_0 magnetic field. After 5^*T_1, over 99% of the spins are aligned. Therefore, if $T_1 = 2000$ ms, then equilibrium is established in about 10 seconds.

In NMR, the equilibrium magnetisation is excited to generate a transverse component, which produces the NMR signal in the RF receiver coil. If a 90° RF pulse is used, then all of the magnetisation is *flipped* into the transverse plane and the spin system is not yet ready for a new excitation pulse. The transverse magnetisation must give up its energy to the local environment in order to re-establish the longitudinal (equilibrium) magnetisation, which is aligned along the Z-axis. This process is characterised by T_1 and is also defined by:

$$M(z) = M_o\left(1 - \exp\left(\frac{-TR}{T_1}\right)\right)$$

where $M(z)$ is the recovering longitudinal magnetisation, M_o is the original equilibrium magnetisation and TR is the time after the 90° RF pulse (see Figure 9.11).

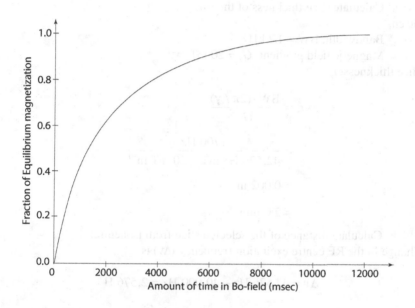

FIGURE 9.11 T_1 magnetisation.

9.14 SLICE THICKNESS

Problem:
A magnetic field gradient of 20 mT m⁻¹ is combined with a static magnetic field of 1 Tesla. A crafted radio frequency pulse with a centre frequency of 42,576 Hz excites a bandwidth of 1.7 kHz.

a. If the magnetic field gradient is turned on while the radio frequency is pulsed, what is the resulting slice thickness of the image?

b. If the central frequency $\left(\dfrac{\omega_o}{2\pi}\right)$ of the crafted sinc pulse is changed to 40,000 Hz, how far will the selected slice be from isocentre?

(Let $\dfrac{\gamma}{2\pi} = 42{,}576 \text{ Hz mT}^{-1}$)

Solution:

a. Calculate slice thickness of the image:

Given,

Bandwidth, BW = 1.7 kHz

Magnetic field gradient, $G_z = 20 \text{ mT m}^{-1}$

Slice thickness, t_s

$$t_s = \frac{\text{BW} \cdot (2\pi / \gamma)}{G_z}$$

$$= \frac{1700 \text{ Hz}}{42{,}576 \text{ Hz mT}^{-1} \ 20 \text{ mT m}^{-1}}$$

$$= 0.002 \text{ m}$$

$$= 2.0 \text{ mm}$$

b. Calculate distance of the selected slice from isocentre:

Change in the RF centre excitation frequency (Δv) is

$$\Delta v = 42{,}576 \text{ Hz} - 40{,}000 \text{ Hz} = 2{,}576 \text{ H}$$

Hence, the distance of the selected slice from the isocentre would be,

$$\Delta z = \frac{\Delta v \cdot (2\pi / \gamma)}{G_z}$$

$$= \frac{2576 \text{ Hz}}{42{,}576 \text{ Hz mT}^{-1} 20 \text{ mT m}^{-1}}$$

$$= 0.003 \text{ m}$$

$$= 3.0 \text{ mm}$$

The thickness of an excited slice is a function of two factors. First is the duration of the RF pulse. The longer the RF pulse, the narrower the transmitter

(TX) bandwidth (BW) and therefore the narrower the slice is. However, there is a limit to the amount of time that an RF pulse should play out; therefore, it is unusual to have a selected slice that is less than 2 mm. The other factor that controls slice thickness is the amplitude of the slice selection gradient. The amplitude of the slice selection gradient is proportional to the transmitter bandwidth and inversely proportional to the product of the prescribed slice thickness and the reduced gyromagnetic ratio ($2\pi/\gamma$). The change in the RF centre excitation frequency (Δv) allows multiple slices to be acquired at different slice locations.

9.15 FIELD OF VIEW

Problem:
An x-gradient coil is used with the MRI system for frequency encoding. What will the field of view (FOV_x) in the frequency-encoding direction be if the x-gradient coil is turned on with a $G_x = 3$ mT m^{-1} during signal sampling with a receiver bandwidth of $\text{BW}_{RX} = 32$ kHz for protons $\left(\dfrac{\gamma}{2\pi} = 42{,}576 \text{ Hz mT}^{-1} \right)$?

Solution:

$$\text{FOV}_x = \frac{2\pi}{\gamma} \cdot \frac{\text{BW}_{RX}}{G_x}$$

$$= \frac{32{,}000 \text{ Hz}}{\left(42{,}576 \text{ Hz mT}^{-1} \times 3 \text{ mT m}^{-1} \right)}$$

$$= 0.2505 \text{ m}$$

$$= 25 \text{ cm}$$

There is an inverse relationship between the strength of the frequency-encoding gradient and the size of the field of view in the frequency encoding direction (FOV_x). The FOV_x is proportional to the receiver bandwidth divided by the product of the gyromagnetic ratio and the frequency-encoding gradient strength. The receiver bandwidth is completely different from the RF or transmitter bandwidth (BW_{TX}). The receiver bandwidth is proportional to the FOV_x and the frequency encoding gradient strength, G_x, is inversely proportional to FOV_x. The selection of the receiver bandwidth has important implications for imaging time, geometric distortion artefacts and the overall signal-to-noise ratio (SNR) of the MR image.

Problem:

A y-gradient is turned on following successive excitations but before 256 signal detections. Furthermore, the y-gradient is incremented by an amount ΔG_y = 0.0375 mT m^{-1} between each NMR signal sampling for a duration t_y = 5 ms. (depicted in Figure 9.12). These data can be used to produce an image with what field of view (FOV$_y$) in the phase-encoding, y-direction?

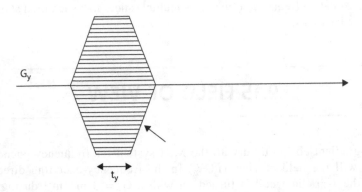

FIGURE 9.12   MRI Phase-encoding gradient waveform. The multiple horizontal lines indicate incremental increases in the phase-encoding gradient amplitude for successive signal acquisitions.

Solution:

$$\text{FOV}_y = \frac{\pi / \gamma}{(\Delta G_y \cdot t_y)}$$

$$= \frac{1}{(85,152 \text{ Hz mT}^{-1} \cdot 0.0375 \text{ mT m}^{-1} \cdot 0.005 \text{ s})}$$

$$= 0.0626 \text{ m}$$

$$= 6.26 \text{ cm}$$

In a similar manner to the slice-select gradient and the frequency-encoding gradient, there is an inverse relationship between the phase-encoding gradient and the field of view (FOV$_y$) in the phase encoding direction. In this case, the value of the matrix size in the phase encoding direction (N_y = 256) divided by two times the duration of the phase encoding gradient (T_y) takes the role of the bandwidth. Note that there is no factor of two in this equation because the phase encoding gradient runs from a maximum negative value to a maximum positive value. $N_y/2 \times$ 0.0375 mT m^{-1} = 4.8 mT m^{-1}, which is the maximum value of G_y. Another way to view this relationship is that the phase-encoding gradient increment (ΔG_{ph}) is equal to π divided by the product of the gyromagnetic ratio (γ), the FOV_y and T_y.

9.16 SIGNAL TO NOISE RATIO

Problem:
An MR image is constructed with slice thickness of 5 mm, FOV = 25 cm in both the phase-encoding and frequency-encoding directions, with a 256 × 256 matrix. The receiver bandwidth, BW_{RX1} = 32 kHz and the signal-to-noise ratio (SNR_1) is measured to be 100. A second image is then acquired with all parameters the same except that the BW_{RX2} = 16,000 Hz. What is the SNR_2 of the new image?

Solution:
The SNR is inversely proportional to the square root of BW_{RX}. Therefore, the new SNR:

$$SNR_2 = SNR_1 \sqrt{\left(\frac{BW_{RX1}}{BW_{RX2}}\right)}$$

$$= 100 \sqrt{\left(\frac{32,000}{16,000}\right)}$$

$$= 141$$

In general, the SNR of an MR image is related to the image parameters by the following proportionality:

$$SNR \propto M_{xy} \cdot t_s \cdot \frac{FOV_x}{N_x} \cdot \frac{FOV_y}{N_y} \cdot \frac{\sqrt{NSA}}{BW_{RX}}$$

Where,

M_{xy} = net transverse magnetisation (controlled by a number of parameters including γ, T_1, T_2, TR, and TE,...),

t_s = slice thickness,

$\dfrac{FOV_x}{N_x}$ = pixel size in the frequency encoding direction,

$\dfrac{FOV_y}{N_y}$ = pixel size in the phase-encoding direction,

NSA = number of signals averaged, and BW_{RX} = receiver bandwidth.

9.17 SPIN ECHO PULSE

Problem:
Determine the relative signal amplitudes of white matter (WM) versus cerebrospinal fluid (CSF) for a spin echo pulse sequence with TR = 500 ms and TE = 30 ms. Proton density, T_1 and T_2 values for CSF and WM at B_o = 1.5 T are shown in the Table 9.1.

TABLE 9.1 Proton density, T_1 and T_2 properties of brain matter

TISSUE	PROTON DENSITY (ρ, % OF WATER)	T_1 (ms)	T_2 (ms)
Grey matter	83%	900	100
White matter	70%	780	90
CSF	100%	2400	160

Solution:
The signal intensity derived from a spin echo pulse sequence is modulated by selection of the TR and TE timing parameters. By changing TE and TR, we can get different contrast weighting. The signal can be represented as:

$$S_{(\text{TE},\text{TR})} = S_o \left(1 - e^{-\text{TR}/T_1} \right) \times e^{-\text{TE}/T_2}$$

Where,

S_o is the signal in the limit that TE→0 and TR → infinity,
TR is the repetition time,
TE is the time to echo, and
T_1 and T_2 denote relaxation properties of the tissues.

This soft tissue contrast can be controlled by optimising the timing parameters in MRI pulse sequences.

$$S_{\text{CSF}} = k\, \rho_{\text{CSF}} \left[1 - \exp\left(-\frac{\text{TR}}{T_{1\text{CSF}}} \right) \right] \times \exp\left(\frac{\text{TE}}{T_{2\text{CSF}}} \right)$$

$$= k \cdot 1.0 \cdot \left[1 - \exp\left(-\frac{500 \text{ ms}}{2400 \text{ ms}} \right) \right] \times \exp\left(-\frac{30 \text{ ms}}{160 \text{ ms}} \right)$$

$$= k \cdot 1.0 \cdot 0.188 \cdot 0.829$$

$$= 0.156$$

$$S_{WM} = k\ \rho_{WM} \left[1 - \exp\left(-\frac{TR}{T_{1WM}} \right) \right] \times \exp\left(-\frac{TE}{T_{2WM}} \right)$$

$$= k \cdot 0.7 \cdot \left[1 - \exp\left(-\frac{500}{780} \right) \right] \times \exp\left(-\frac{30}{90} \right)$$

$$= k \cdot 0.7 \cdot 0.473 \cdot 0.717$$

$$= 0.237$$

$$\frac{S_{WM}}{S_{CSF}} = 1.52$$

Note:
k is the instrumental scaling constant.

9.18 INVERSION RECOVERY

Problem:
In an inversion recovery experiment, data points are collected for two values of TI. For TI = 200 ms the signal amplitude is –22. For TI = 300 ms the signal amplitude is +5.

    a. Estimate the value of TI for which the signal will be zero.
    b. Based on your calculations, determine the T_1 of the sample under study.

Solution:
Inversion recovery uses a 180° inversion preparation pulse to precondition the longitudinal magnetisation before the excitation pulse is applied and image acquisition begins. The inversion pulse is followed by a normal spin echo acquisition. The application of the inversion pulse introduces a new timing parameter, TI, which is the time between the inversion pulse and the excitation pulse. Now, TR is longer than the sum of TI and TE. Inversion of the longitudinal magnetisation effectively doubles the dynamic range of the T_1 relaxation process. The recovery equation, in the case of inversion, shows that magnetisation is equal to the equilibrium magnetisation, $M_o \times (1 - 2 \times \exp[-TI/T_1])$.

    The longitudinal magnetisation recovers from a maximum negative value to the maximum positive value, in the process crossing through a zero value.

a. Estimate the value of TI for which the signal will be zero:

Given,

TI (ms)	SIGNAL (AU)
200	−22
300	5

Interpolate for TI(0) where S = 0:

$$TI(0) = 200 + \left[\frac{(300 - 200)}{(5 - (-22))} \right] \cdot 22 = 281.5 \text{ ms}$$

b. Solving for T_1:

If $S = S_0 \left[1 - 2\exp\left(-\frac{TI}{T_1} \right) \right]$,

then $\dfrac{S}{S_0} = 1 - 2\exp\left(-\frac{TI}{T_1} \right)$,

or $\left(1 - \dfrac{S}{S_0} \right) = 2\exp\left(-\frac{TI}{T_1} \right)$.

Then, $\dfrac{\left(1 - \dfrac{S}{S_0} \right)}{2} = \exp\left(-\frac{TI}{T_1} \right)$.

So, $-\dfrac{TI}{T_1} = \ln\left[\dfrac{\left(1 - \dfrac{S}{S_0} \right)}{2} \right]$.

If $S = 0$, then

$$-\frac{TI(0)}{T_1} = \ln\left[\frac{1}{2} \right]$$

$$-\frac{TI(0)}{T_1} = -0.693$$

$$T_1 = 1.44 \, TI(0)$$

$$T_1 = 1.44 \times 281.5 \text{ ms} = 405.4 \text{ ms}$$

9.19 ACQUISITION TIME

Problem:

A standard multi-slice T_2-weighted spin echo has a matrix size of 192(pe) × 256 (fe), TR = 2200 ms, TE = 80 ms, signal averages: Number of signal average (NSA) = 1, 20 slices. How long does the scan require for completing data acquisition?

[pe = phase encoding, fe = frequency encoding]

Solution:

If TE = 80 ms and TR = 2200 ms, there should be enough time to acquire signal from all 20 slices for a single phase-encoding value within a single TR period.

Then, the total acquisition time is: $T_{acq} = NSA \times TR \times N_{pe}$

$$T_{acq} = 1 \times 2.2 \text{ s} \times 192 = 422.4 \text{ s} = 7.04 \text{ min}$$

The total scan time in a spin echo pulse sequence is determined by three factors. First the TR is selected to obtain the proper contrast for a given pulse sequence. Second, the number phase encoding steps (N_{pe}) is determined by the matrix size required to obtain adequate spatial resolution. Finally, number of signals co-added (NSA) is determined by the need for adequate signal-to-noise ratio. SNR can be improved by using higher B_0 magnets and enhancing the sensitivity of the MRI radiofrequency receiver coils. However, the number of phase-encoding views can be optimised for specific clinical situations.

9.20 CHEMICAL SHIFT ARTEFACT

Problem:

If the kidneys are imaged in a 3 T magnet using a spin echo pulse sequence with FOV = 25 cm, matrix = 256 × 256, BW = 32 kHz. (fat chemical shift = 3.4 ppm)

a. How many pixels will the fat-water shift be in the phase-encoding and frequency-encoding directions?

b. In an EPI sequence with same parameters except matrix = 100 × 100 and BW = 500,000 Hz, how many pixels will the fat-water shift be in the frequency-encoding directions?

Solution:

a. SE sequence:

Receiver bandwidth (in Hz pixel^{-1}) = $\dfrac{32,000}{256}$ = 125 Hz pixel^{-1}

Chemical shift = 3.4 ppm × 3T × 42.576 Hz ppm^{-1} T^{-1} = 434.3 Hz

Frequency-encode F-W shift = $\dfrac{434.3 \text{ Hz}}{125 \text{ Hz pixel}^{-1}}$ = 3.47 pixels

Phase-encode F-W shift = 0 pixels

b. EPI sequence:

Chemical shift = 3.4 ppm × 3T × 42.576 Hz ppm^{-1} T^{-1} = 434.3 Hz

Receiver bandwidth (in Hz pixel^{-1}) = $\dfrac{500,000}{100}$ = 5,000 Hz pixel^{-1}

Frequency-encode F-W shift = $\dfrac{434.3 \text{ Hz}}{5,000 \text{ Hz pixel}^{-1}}$ = 0.087 pixels

Phase-encode dwell time = $\dfrac{1}{500 \text{ kHz}}$ = 2 μs

Therefore, time between phase-encoding samples = 2 μs/sample × 100 samples = 200 μs.

Effective BW$_{PE}$ = $\dfrac{1}{200 \text{ s}}$ = 5,000 Hz; $\dfrac{5,000 \text{ Hz}}{100 \text{ pixels}}$ = 50 Hz pixel^{-1}

Phase-encode F-W shift = $\dfrac{434.3 \text{ Hz}}{50 \text{ Hz pixel}^{-1}}$ = 8.7 pixels

The chemical shift artefact (CSA) typically appears in the frequency encoding direction; it will appear in the phase encoding direction for high-speed, echo planar imaging. An anatomical location at which the CSA is often observed is the kidney, which is encapsulated within a protective layer of retroperitoneal fat. In addition to increasing the receiver bandwidth, the chemical shift artefact can be enhanced using fat selective RF saturation and/or a short TI inversion recovery sequence (STIR) with the inversion time set to the crossing point of fat. The CSA does not appear in the phase-encoding direction in SE images because the phase is reset by having a new RF excitation pulse for each phase-encoding step. In EPI, the CSA exists but is negligible in the frequency-encoding direction.

9.21 SPECIFIC ABSORPTION RATE

Problem:

The MRI system calculates average specific absorption rate (SAR) = 1.3 W kg^{-1} averaged over the head for a 10-minute spin-echo pulse sequence.

a. Explain what SAR in MRI is and the factors that would affect the SAR.

b. You decide to increase the number of slices from 20 to 30 and reduce the TR from 2000 ms to 1333 ms in order to accomplish this in the same scan time. What is the new SAR?

c. You do not like the contrast at TR= 1333s and decide to use parallel imaging (SENSE) with $R = 2$ and TR = 2000 ms with your 30-slice acquisition. What is the SAR now?

Solution:

a. The patient is in an RF magnetic field that causes spin excitation (the $B1$ field). The RF field can induce small currents in the electrically conductive patient, which result in energy being absorbed. The RF power absorbed by the body is called the specific absorption rate (SAR). SAR has units of watts absorbed per kg of patient. If the SAR exceeds the thermal regulation capacity, the patient's body temperature will rise. RF pulses deposit heat into tissues and vary depending on the type of RF pulse (i.e. 90° or 180°), the number of RF pulses, the pulse widths, TR, size of the patient and type of transmitter coil.

 • Patient size: SAR increases as the patient size increases – directly related to patient radius.

 • Resonant frequency: SAR increases with the square of the Larmor frequency.

 • RF pulse flip angle: SAR increases as the square of the flip angle.

 • Number of RF pulses: SAR increases with the number of RF pulses in a given time.

b. The SAR is directly proportional to the duty cycle, defined as the fraction of total scan time for which the RF is present.

By increasing the number of slices from 20 to 30 and reducing the TR from 2000 ms to 1333 ms, the duty cycle is increased by a factor of $\left(\dfrac{30}{20}\right) \times \left(\dfrac{2000}{1333}\right) = 2.25$, so the new SAR = 2.25×1.3 W kg^{-1} = 3.0 W kg^{-1}.

c. By keeping the same TR but increasing the number of slices and reducing the number of phase-encoding steps by ½ using SENSE, the duty cycle changes by a factor of $\left(\dfrac{30}{20}\right) \times \left(\dfrac{1}{2}\right) = 0.75$, so the SAR is reduced to 0.75×1.3 W kg^{-1} = 0.98 W kg^{-1}.

However, the SNR is also reduced.

Note:
Theoretically, parallel MR imaging (pMRI) methods like SENSE can be used to reduce scan time by a factor (R-factor), which can be as high as the number of phased-array coil elements. However, if full imaging acceleration is attempted, numerical instabilities will result, which quickly degrade the SNR. Typically, decreased SNR and increased image non-uniformities limit practical use of acceleration factors well below the theoretical maximum R. In parallel imaging, the number, size and orientation of the coil elements influence the SNR in an unusual manner compared to normal imaging. The 'geometric factor', g, is the coil-dependent noise amplification factor across the image volume that comes from the estimation process of the unwrapping algorithm. It can be measured by determining the SNR in the pMRI data compared to the SNR obtained under normal conditions. The geometry factor is a function of both spatial position and R, increasing as $1/\sqrt{R}$. The minimum g-factor is unity but good g-factors range between 1.0 and 1.5.

Ultrasound

10

10.1 AXIAL RESOLUTION AND LATERAL RESOLUTION

Problem:

Explain briefly the terms 'axial resolution' and 'lateral resolution' in ultrasound imaging. Describe briefly how axial and lateral resolution vary with transducer frequency and penetration depth.

Solution:

Figure 10.1 shows a diagram explaining axial and lateral resolution in ultrasound imaging.

Axial resolution

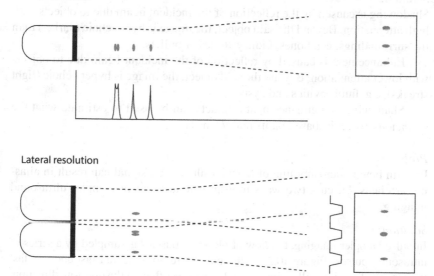

Lateral resolution

FIGURE 10.1 Axial and lateral resolutions.

Axial resolution is the resolution as measured along the beam path.

Lateral resolution is the resolution as measured perpendicularly in reference to the beam path.

Lateral resolution improves by increasing the frequency while axial resolution is inversely proportional to the frequency of the transducer depending on the size of the patient.

Lateral resolution varies with the depth (due to the shape of the beam). Lateral resolution is best at the focal zone.

Axial resolution is inversely proportional to the frequency of the transducer depending on the size of the patient. The higher the frequency the lower the axial resolution is in large patients. This state results from the rapid absorption of the ultrasound energy with lower penetration. Lower frequencies are utilized to increase depth of penetration.

10.2 ARTEFACTS

Problem:

Explain briefly what causes shadowing and enhancement artefacts in an ultrasound image, and state examples of materials that may cause this. Finally, state a situation where these artefacts can be diagnostically useful.

Solution:

Shadowing is caused by the reflection of the incident beam due to objects with high attenuation. Beyond the said object, the image is hypoechoic (darker) than the surroundings, e.g. bones, kidney stones or bullets.

Enhancement is caused by reflection of the incident beam due to objects with lower attenuation. Beyond the said object, the image is hyper echoic (light streaks), e.g. fluid cavities and cysts.

Shadowing and enhancement artefacts can be used to estimate what the suspicious object is, based on its mass/density.

Problem:

Explain how under-sampling of Doppler ultrasound signal can result in aliasing artefacts. Discuss two ways to eliminate aliasing in Doppler ultrasound imaging.

Solution:

In pulse Doppler imaging, the flow of blood corpuscles is sampled by a series of ultrasound pulses (Figure 10.2). If the pulses are not repeated fast enough, fast flow in one direction will be interpreted as a slower flow in the opposite direction.

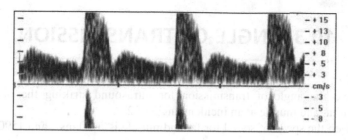

FIGURE 10.2 An example of aliasing, where high velocity blood flow in Doppler ultrasound 'wraps' around and shows up as lower blood flow velocity.

One method to avoid this artefact is to use continuous wave imaging. There is no limitation in pulse repetition frequency, unlike pulsed Doppler. Alternatively, one can use a lower frequency transducer. The Doppler shift is related to the frequency of the transducer and, a lower frequency will produce smaller Doppler shifts.

Problem:
Name and describe the ultrasound artefact depicted in Figure 10.3. Give a common situation where this artefact might occur.

Solution:
Reverberation artefact arises from multiple echoes generated between two closely spaced interfaces that are highly reflective, reflecting ultrasound energy back and forth during the acquisition of the signal and before the next pulse. This artefact can be seen when imaging air pockets in the anatomy, e.g. bowel gasses. In this case, this would be due to multiple reflections between the transducer and the anterior bladder wall.

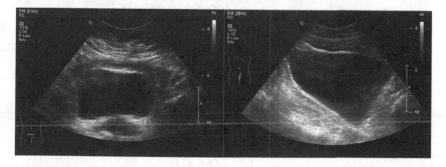

FIGURE 10.3 Ultrasound artefact.

10.3 ANGLE OF TRANSMISSION

Problem:
Calculate the angle of transmission for ultrasound striking the interface between fat and muscle at an incident angle of 25°.

Given the speed of sound for fat and muscle is 1450 ms⁻¹ and 1590 ms⁻¹, respectively.

Solution:
According to Snell's Law,

$$\frac{\sin\theta_i}{\sin\theta_t} = \frac{1450}{1590}$$

$$\frac{\sin 25°}{\sin\theta_t} = \frac{1450}{1590}$$

Angle of transmission, $\theta_t = 27.6°$

10.4 REFLECTION COEFFICIENTS

Problem:
Calculate the intensity reflection coefficients and the intensity transmission coefficients when an ultrasound beam encounters the following tissue interfaces at an incident angle of 90°. Explain the three conditions where the acoustic impedance, Z of the tissues at the interface are $Z_1 \gg Z_2$, $Z_1 \sim Z_2$, and $Z_1 < Z_2$.

	ACOUSTIC IMPEDANCE, Z	
Case	Z_1, (g cm⁻²s⁻¹)	Z_2, (g cm⁻²s⁻¹)
$Z_1 \gg Z_2$	Muscle, (1.7)	Air, (0.00043)
$Z_1 \sim Z_2$	Liver, (1.65)	Kidney, (1.62)
$Z_1 < Z_2$	Muscle, (1.7)	Bone, (7.8)

Solution:
Intensity reflection coefficient,

$$R = \frac{I_r}{I_i} = \frac{(Z_2 - Z_1)^2}{(Z_2 + Z_1)^2}$$

Intensity transmission coefficient,

$$T = \frac{I_t}{I_i} = 1 - R = \frac{4(Z_1 Z_2)}{(Z_2 + Z_1)^2}$$

Therefore,

	Z_1, (g cm^{-2}s^{-1})	Z_2, (g cm^{-2}s^{-1})	R (%)	T (%)
$Z_1 \gg Z_2$	Muscle, (1.7)	Air, (0.00043)	99.90	0.10
$Z_1 \sim Z_2$	Liver, (1.65)	Kidney, (1.62)	0.01	99.99
$Z_1 < Z_2$	Muscle, (1.7)	Bone, (7.8)	41.23	58.77

When $Z_1 \gg Z_2$, ultrasound travels from muscle to air cavity, e.g. in the GI tract imaging. 99.9% of the ultrasound signal is reflected in front of the air cavity, and only a little fraction of ultrasound signal is transmitted through. Therefore, no information can be obtained from the structures behind the air cavity. This condition also describes the interface of tiny air layer between the ultrasound probe and patient's skin. Therefore, coupling gel needs to be applied onto the patient's skin surface to exclude the thin layer of air.

When $Z_1 \sim Z_2$, ultrasound traverses tissues of similar acoustic impedance. Almost 100% of the signal will be transmitted through with very little (0.01%) or no backscattered/reflected ultrasound signal. The tissue boundary is essentially undetectable or invisible to the ultrasound signal.

When $Z_1 < Z_2$, ultrasound traverses tissues with dissimilar acoustic impedances but yet not too different; the proportion of reflected and transmitted ultrasound signals are in almost equal proportions.

10.5 DOPPLER ULTRASOUND

Problem:

   a. Calculate the Doppler frequency, f_D, given
      Incident ultrasound frequency, $f_i = 5.5$ MHz
      Probe angle, $\theta = 45°$
      Blood velocity, $v = 0.4$ m s^{-1}
      Speed of sound of muscle, $c = 1590$ m s^{-1}.

   b. Explain what happens if the probe angle increases.
   c. In practice, what is the optimum angle of the probe?

Solution:

  a. Doppler frequency, f_D

$$f_D = \frac{2 f_i v \cos \theta}{c}$$

$$= \frac{2(5.5 \times 10^6)(0.4) \cos 45°}{1590}$$

$$= 1.96 \text{ kHz}$$

The Doppler frequency is 1.96 kHz, which is within the audible sound range. That is why the Doppler ultrasound is usually accompanied by a pulsating sound.
  b. As the probe angle increases, the Doppler frequency reduces and the pulsating sound becomes softer.
  c. The accuracy of Doppler measurement is determined by the knowledge of the probe angle. The blood velocity measurement error is least when the probe angle is small. An error of 3° in the angle probe will result in more than 40% error in the measurement when the probe angle is > 60°.

10.6 ULTRASOUND TRANSDUCER

Problem:
What is the purpose of the backing layer and the impedance-matching layer in a typical ultrasound transducer?

Solution:
The backing layer is coupled to the piezoelectric material. It absorbs the backward directed ultrasound energy and eliminates any echoes that could potentially return to the transducer from the housing assembly. It is a damping device designed to restrict the time of vibration and keep the ultrasound pulse short in order to preserve details along the beam axis (axial resolution). If there is no mechanical damping, the piezoelectric material will 'ring' after the end of the voltage pulse, producing a pressure wave that is longer than the voltage pulse.

The impedance matching layer provides an interface between the transducer element and the tissue, minimising the acoustic impedance differences between the transducer and the patient. The purpose of this is to avoid creating a large difference of characteristic acoustic impedance between the two adjacent materials, which would result in a large amount of reflected signal due to poor coupling of signal.

10.7 ATTENUATION

Problem:
The intensity of a 3 MHz ultrasound beam entering tissue is 10 mWcm^{-2}. Calculate the intensity at a depth of 4 cm. (The attenuation coefficient is 1 dB cm^{-1} MHz^{-1}.)

Solution:
The attenuation coefficient is 1 dB cm^{-1} MHz^{-1}, and so has a value of 3 dB cm^{-1} at 3 MHz. At a depth of 4 cm, the attenuation is 12 dB, which corresponds to a factor of 16. So the intensity of the beam is 0.625 mWcm^{-2}.

$$\text{attenuation} = 10\log_{10}\frac{I_1}{I_o}\,\text{dB}$$

$$-12 = 10\log_{10}\frac{I_1}{I_o}$$

$$-1.2 = \log_{10}\frac{I_1}{I_o}$$

$$10^{-1.2} = \frac{I_1}{I_o}$$

$$10^{-1.2}I_o = I_1$$

$$I_1 = 10^{-1.2}\left(10\,\text{mWcm}^{-2}\right)$$

$$I_1 = 0.625\,\text{mWcm}^{-2}$$

10.8 ULTRASOUND ATTENUATION

Problem:
A 5 MHz ultrasound beam incident perpendicularly onto a patient body, traversing 2 cm of muscle tissue, 3 cm of fat and 4 cm of liver. The tissue properties are given in Table 10.1.

a. Calculate the reflection index of the signal at the two interfaces.
b. Calculate the total energy loss of the ultrasound signal.

Give a conclusion to your calculation.

TABLE 10.1   Ultrasound properties for some human tissues

TISSUE TYPE	CHARACTERISTICS ACOUSTIC IMPEDANCE, Z (kg m^{-2} s^{-1} or rayls)	ATTENUATION COEFFICIENT, α (dB cm^{-1} MHz^{-1})
Muscle	1.70×10^{-6}	0.6
Fat	1.38×10^{-6}	1.0
Liver	1.60×10^{-6}	0.4

Solution:

a. The reflection index is calculated using the following equation:

$$R = \frac{I_r}{I_i} = \left[\frac{Z_2 - Z_1}{Z_2 + Z_1}\right]^2$$

For the muscle – fat interface:

$$R_{\text{muscle-fat}} = \frac{I_r}{I_i} = \left[\frac{Z_{\text{fat}} - Z_{\text{muscle}}}{Z_{\text{fat}} + Z_{\text{muscle}}}\right]^2 = \left[\frac{1.38 - 1.7}{1.38 + 1.7}\right]^2 = 0.01$$

Only 1% of the beam signal is reflected.
For the fat – liver interface:

$$R_{\text{fat-liver}} = \frac{I_r}{I_i} = \left[\frac{Z_{\text{liver}} - Z_{\text{fat}}}{Z_{\text{liver}} + Z_{\text{fat}}}\right]^2 = \left[\frac{1.6 - 1.38}{1.6 + 1.38}\right]^2 = 0.0055$$

Only 0.5% of the beam signal is reflected.
b. The total energy loss is calculated as

Total energy loss $=$ (Energy loss in muscle) $+$ (Energy loss in fat)

$+$ (Energy loss in liver)

$= (0.6 \text{ dB cm}^{-1} \text{ MHz}^{-1} \times 2 \text{ cm} \times 5 \text{ MHz})$

$+ (1.0 \text{ dB cm}^{-1} \text{ MHz}^{-1} \times 3 \text{ cm} \times 5 \text{ MHz})$

$+ (0.4 \text{ dB cm}^{-1} \text{ MHz}^{-1} \times 4 \text{ cm} \times 5 \text{ MHz})$

$= 29 \text{ dB}$

For an ultrasound signal that traverses these tissue layers and back, the total attenuation is twice that of 29 dB, i.e. 58 dB.

10.9 Q FACTOR

Problem:

Define the 'Q factor' in medical diagnostic ultrasound by its equation. Given that all other factors remain the same, describe how axial resolution is affected by Q factor.

Solution:

The Q Factor is given by $Q = \dfrac{f_0}{\text{Bandwidth}}$ where f_0 is the centre frequency of the sound emanating from the transducer, and bandwidth is the width of the frequency distribution. A low Q factor, with heavy damping and broad bandwidth, results in shorter pulse lengths that produces better axial resolution.

10.10 CONTRAST

Problem:

Explain the physics behind the use of contrast agents in ultrasound. What physical property and type of interaction does a contrast-enhanced ultrasound make use of? State one example of an element which can be used as a contrast agent in ultrasound imaging.

Solution:

Contrast agents in ultrasound work by increasing the acoustic contrast difference between the agent and the tissues. The major interaction between tissues that affect contrast is reflection, and the physical property that largely affects reflection between the tissues is acoustic impedance. Air has a very small acoustic impedance value compared to soft tissue and is commonly used as a contrast agent (in the form of microbubbles) in ultrasound.

10.11 TRANSDUCER SELECTION

Problem:
A doctor uses 3.5 MHz, 10 mm diameter transducer to visualise a deep-seated tumour at about 15 cm from the surface. He complains about poor visualisation of the tumour. Explain why.

Solution:
Lateral resolution (the ability of the system to resolve objects in a direction perpendicular to the beam direction) is best at the end of the near field.

Given speed of sound in soft tissue, $v = 1540$ ms^{-1}

$$\text{Length of near field} = \frac{d^2}{4\lambda}$$

$$= \frac{d^2 f}{4v}$$

$$= \frac{(0.01 \text{ m})^2 \cdot (3.5 \times 10^6 \text{s}^{-1})}{4 \times 1540 \text{ ms}^{-1}}$$

$$= 0.0568 \text{ m}$$

$$= 5.68 \text{ cm}$$

The focal point of the ultrasound probe is at 5.68 cm, which is much shallower compared to the location of the tumour (at 15 cm from the surface), resulting in poor visualisation of the tumour.

10.12 ULTRASOUND SAFETY

Problem:
Which ultrasound mode has the greatest potential for thermal hazard, and why? What steps should be taken to ensure the prudent use of ultrasound?

Solution:
Spectral Doppler has the greatest potential for thermal hazard. It could potentially produce up to 8°C increase in temperature because of the high pulse repetition frequencies and pulse length. This may be acerbated if the beam is held stationary for a long time.

Steps that can be taken to reduce the probability of thermal hazards in ultrasound includes:

- Monitoring of the thermal index.
- Avoiding scans where there is no medical justification.
- The sonographer should be well trained to use the equipment, scanning technique and interpretation of ultrasound images.
- Keeping up to date on the ultrasound safety guidelines.
- Reducing scan time.
- Ensuring optimum operation of scanner via periodic quality assurance of the ultrasound equipment.
- Keeping output power as low as possible and using overall gain to increase signal.
- Avoiding holding probe stationary for long time.
- Freezing image rather than holding an emitting probe stationary.

Radiation Protection and Radiobiology

11

11.1 IONISATION CHAMBER

Problem:

Draw a labelled diagram of a (free air) ionisation chamber and explain the principle of its operation.

Solution:

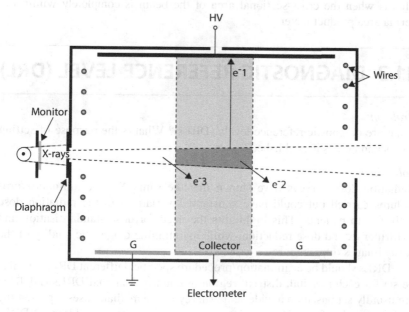

FIGURE 11.1 Free air ionisation chamber.

Free air ionisation chamber (Figure 11.1) consists of parallel-plate or coaxial electrodes in a volume occupied by air. It is enclosed in a Pb shielding box to exclude X-rays scattering arriving from elsewhere. At the front of the box is a tungsten-alloy diaphragm that passes the X-ray beam into the chamber. The entrance of radiation into the chamber results in an electrical current produced as ion pairs are collected by the electrodes. Electrons collected by the anode of an ionisation chamber constitute a direct current. The charge collected from the ionisation chamber may be measured with an electrometer. A dynamic capacitor converts the signal from the ionisation chamber into an alternating current that may be amplified with an AC amplifier and measured. In the ionisation region, the number of primary ions of either sign collected is proportional to the energy deposited by the charged particle tracks in the detector volume.

11.2 KERMA AREA PRODUCT

Problem:
Under what condition will the readings on a kerma area product meter not vary with distance from the tube focus?

Solution:
This is when the cross-sectional area of the beam is completely within the kerma area product meter.

11.3 DIAGNOSTIC REFERENCE LEVEL (DRL)

Problem:
What are diagnostic reference levels (DRLs)? What is the purpose of setting DRLs and how is it established?

Solution:
Radiation dose surveys have shown that the same X-ray examination/procedures carried out could have substantial variations in the radiation dose delivered to patients. This highlights the need for dose standardisation and to further extend dose reduction, while maintaining diagnostic quality of the X-ray images or procedures.

DRLs should be examination/procedure specific. Different DRLs can also be set for each hospital, district, region or country. National DRLs (NDRLs) are usually set based on a wide scale survey of the median doses representing typical practice for specific patient groups (e.g. adults or children). NDRLs are usually set at the 3rd quartile value of the national distributions. Where no national or regional DRLs are available, DRLs can be set based on local dosimetry or practice data, or can be based on published values that are appropriate for the local circumstances.

11.4 SHIELDING 1

Problem:
Discuss the differences between the utility of room shielding as applied for a CT room versus that for a MRI room.

Solution:
Room shielding in a CT room is designed to keep ionising radiation from exiting the room, whereas MRI shielding is designed to prevent external radiofrequencies from interfering with the radiofrequency signal from the nuclear magnetization.

CT room shield will utilise high-Z materials such as lead for its high attenuating properties, whereas MRI shielding utilises highly conductive metallic materials such as copper to produce a Faraday cage that shields the MRI system from external radiofrequency emissions.

11.5 PERSONAL DOSIMETRY

Problem:
What are the advantages and disadvantages of a film badge personal dosimetry system when compared with a thermoluminescent system?

Solution:
Film badge dosimetry
Advantages:

- Permanent record
- Cheap

Disadvantages:

- Require wet chemical processing
- Sensitive to environment
- Not reusable

Thermoluminescent dosimetry:

Advantages:

- Small size
- Reusable
- Can measure a wide range of dose

Disadvantages:

- Lack of uniformity.
- The readout process involves heating and hence erasing the information stored.

11.6 EFFECTIVE DOSE

Problem:
Why is it not recommended to use the effective dose to estimate the level detriment for an individual diagnostic exposure?

Solution:
The tissue weighting factors used in calculating effective dose are based on mean values for the general populations having a mix of old and young, of male and female. It is not appropriate to apply effective dose to individual diagnostic exposure because patient specific parameters may be different from the general assumption.

11.7 CELL SURVIVAL

Problem:
A tumour containing 10^8 cells is inactivated by radiation with a mean lethal dose of $D_o = 1.8$ Gy. What is the number of surviving tumour cells after a radiation dose of 45 Gy?

Solution:
Assuming exponential killing,

$$N = N_o e^{-D/D_o}$$

where,

N = Number of surviving tumour cells
N_o = Original number of tumour cells = 10^8

D = Total delivered dose = 45 Gy
D_0 = Mean lethal dose = 1.8 Gy

$$N = N_0 e^{-D/D_0}$$

$$= 10^8 e^{-45/1.8}$$

$$= 1.39 \times 10^{-3} \text{ cells}$$

The surviving tumour cells of $\ll 1$ cell implies that almost no tumour cells survived and the probability of cell viability is very small.

11.8 TISSUE INJURIES

Problem:
What are the most likely tissue injuries that occur in interventional radiology? What arrangements should be made to minimise such effects and to investigate them when they occur?

Solution:
The most likely tissue injuries that could occur in an interventional radiology procedure would be radiation induced skin injuries and cataracts. The threshold dose for radiation induced skin injuries starts at 2 Gy while the threshold dose for radiation induced cataract formation has been reduced to 0.5 Gy (ICRP, 2011).

To minimise incidences of acute radiation injuries during interventional radiological procedures, it is recommended to shorten irradiation time, reduce the use of magnification, maximise use of digital post-processing techniques available, for example, using the last frame hold instead of continuous fluoroscopy, using pulsed fluoroscopy mode instead of continuous fluoroscopy mode, and applying filtration where possible to reduce skin dose due to low energy photons. Adequate training should be provided to the staff, including the radiographers and interventional radiologists in order to improve their techniques and skills. This would lead to more effective and efficient usage of radiation dose and shorter procedure time. Use small field size where possible to reduce large area exposures because large area exposures will increase patient dose and scattered radiation.

When radiation induced injury occurs, investigation can be carried out through various means such as monitoring of radiation dose to patient via kerma-area product (KAP), air kerma from the scanner console, and direct (*in-vivo*) measurements, such as thermoluminescent dosimeters, radiochromic films, etc.

11.9 SHIELDING 2

Problem:

a. Define the term 'Use factor' and 'Occupancy factor'.
b. Why is standard fraction being used in practice? Give one example for each factor.

Solution:

a. Use factor (U) is the fraction of operating time that the beam is directed towards the barrier.
   Occupancy factor (T) is the fraction of time that a particular place is occupied by staff, patients or public.
b. The use factor can be difficult to calculate, so standard fractions can be used for the walls and floor.
   U ranges between 0 and 1 for primary barrier, depending upon whether the primary beam is directed at that particular wall. $U = 1$ for secondary barrier because scattered radiation is always present in all directions.
   Work area, office staff rooms ($T = 1$), corridors ($T = 1/5$), toilets, unattended waiting room ($T = 1/20$).

11.10 OCCUPATIONAL DOSE

Problem:

a. What is the ICRP statement (2011) recommendation for the equivalent dose to the lens of the eye for occupational exposure? What is the threshold in the absorbed dose of the lens?
b. During a typical fluoroscopy procedure, a radiologist wearing a lead apron and no protective eyewear receives an average uniform whole body exposure of 0.2 mSv. How many similar procedures can the radiologist perform in one week?

Solution:

a. The ICRP statement (2011) recommends the dose limit for the equivalent dose to the lens of the eye as '20 mSv per year, averaged over defined periods of 5 years, with no single year exceeding 50 mSv'.

The threshold in absorbed dose for the lens of the eye is considered to be 0.5 Gy.

b. For a maximum dose limit of 50 mSv per year for eye lens dose, assume 50 weeks in a year; therefore, the dose limit is 1 mSv per week.

Since the lead apron will attenuate 95% of the whole body dose, the eye dose is the limiting factor, and will reach 1 mSv after five 0.2 mSv procedures.

Note:

The equivalent dose limit for the lens of the eye for occupational exposure in planned exposure situations was reduced from 150 mSv per year to 20 mSv per year, averaged over defined periods of 5 years, with no annual dose in a single year exceeding 50 mSv.

References

Allisy-Roberts, P. and Williams, J. 2008. *Farr's Physics for Medical Imaging*, Philadelphia, PA: Saunders Ltd.

Burgess, A. E. 1999. The Rose model, revisited. *Journal of Optical Society America A*, 16, 633–646.

Bushberg, J. T., Seibert, J. A., Leidholdt, E. M. J. and Boone, J. M. 2012. *The Essential Physics of Medical Imaging*. Philadelphia, PA: Lippincott Williams & Wilkins.

Dance, D. R., Christofides, S., Maidment, A. D. A., McLean, I. D. and NG, K. H. (eds.) 2014. *Diagnostic Radiology Physics: A Handbook for Teacher and Students*. Vienna, Austria: IAEA.

Dendy, P. P. and Heaton, B. 2012. *Physics for Diagnostic Radiology*, Boca Raton, FL: CRC Press.

IAEA Radiation Protection of Patients (RPOP). Diagnostic Reference Levels (DRLs) for Medical Imaging. Available at: https://rpop.iaea.org/RPOP/RPoP/Content/InformationFor/HealthProfessionals/1_Radiology/Optimization/diagnostic-reference-levels.htm.

ICRP 2012. ICRP Statement on tissue reactions/early and late effects of radiation in normal tissues and organs – Threshold doses for tissue reactions in a radiation protection context. ICRP Publication 118. *Ann. ICRP* 41, 1–2.

Ng, K. H., Wong, J. H. D. and Tan, S. K. 2017. Technical specifications of medical imaging equipment. In Borrás, C. (ed.), *Defining the Medical Imaging Requirements for a Rural Health Center*. Singapore: Springer Singapore.

Norweck, J. T., Seibert, J. A., Andriole, K. P., Clunie, D. A., Curran, B. II., Flynn, M. J., Krupinski, E., Lieto, R. P., Peck, D. J., Mian, T. A. and Wyatt, M. 2013. ACR–AAPM–SIIM technical standard for electronic practice of medical imaging. *Journal of Digital Imaging*, 26, 38–52.

Seibert, J. A. and Morin, R. L. 2011. The standardized exposure index for digital radiography: An opportunity for optimization of radiation dose to the pediatric population. *Pediatric Radiology*, 41, 573–581.

Yaffe, M. J. 1994. In Haus, A. G., Yaffe, M. J., (eds), *Syllabus: A Categorical Course in Physics: Technical Aspects of Breast Imaging*. Oak Brook, IL: RSNA.

Printed in the United States
by Baker & Taylor Publisher Services